Microbial Resources for Crop Improvement

Microbial Resources
for
Crop Improvement

Editors

Bishwanath Chakraborty
Usha Chakraborty

Department of Botany
University of North Bengal
Siliguri-734013
West Bengal, India

SSPH

SATISH SERIAL PUBLISHING HOUSE

403, Express Tower, Commercial Complex, Azadpur, Delhi-110033 (India)
Phone : 011-27672852, Fax : 91-11-27672046
E-mail : info@satishserial.com, hkjain1975@yahoo.com
Website : www.satishserial.com

Published by :

SATISH SERIAL PUBLISHING HOUSE

403, Express Tower, Commercial Complex, Azadpur, Delhi-110033 (INDIA)
Phone : 011-27672852 Fax : 91-11-27672046
E-mail : info@satishserial.com, hkjain1975@yahoo.com

ISBN 978-93-81226-39-1

Typeset at: Harminder Kharb *for* Laxmi Art Creation # 9971212252
Printed at: Salasar Imaging Systems, Delhi-110035

Professor Samir Kumar Das
Ph.D.
Vice Chancellor
University of North Bengal

University of North Bengal
Website : http://www.nbu.ac.in
E-mail : nbuvc@nbu.ac.in

FOREWORD

I am glad to know that the book entitled **Microbial Resources for Crop Improvement** is being published by Satish Serial Publishing House, Delhi, incorporating 21 research articles presented to a National Symposium entitled "Microbial Wealth Plant Health" organized by Department of Botany, University of North Bengal under its DRS Programme. I am sure the publication will add another laurel to Department of Botany, University of North Bengal. I wish the publication all success and extend my warm wishes to all my colleagues.

Samir Kumar Das
Vice Chancellor

PREFACE

Microbes are omnipresent and exert immense influence on the life of all other living organisms, whether it is higher plant or man. Probably, life as we know today would not have been possible without the role played by microorganisms which act as invisible actors. Microbes may be beneficial or harmful to plants or human and hence need constant attention whether to optimize their beneficial activities or minimize their harmful effects.

The present volume is a collection of research findings which deal extensively with the microbe-plant interaction which are either beneficial or harmful. Many papers focus on the utilization of microbes as bio resources while others have focused on how disease causing microorganisms can be detected and managed. These research papers were presented at the national symposium on *Microbial Wealth-Plant Health* organised by the Department of Botany, University of North Bengal. By bringing these papers into print, we hope that researchers, teachers and students of plant pathology, microbiology and biotechnology will be benefited. The articles in this book will also act as catalysts in ushering newer ideas in providing meaningful solutions to intricate problems in plant pathology.

We are thankful to Dr. Pannalal Dey, Post-Doctoral Fellow, Immuno-Phytopathology Laboratory, Department of Botany, University of North Bengal for his untiring help in computational work. We wish to thank our publisher Satish Serial Publishing House, Delhi for cooperation in publishing the book in stipulated time using the latest printing technology.

B.N. Chakraborty
U. Chakraborty

LIST OF CONTRIBURS

A. Ghosh
Plant Biotechnology laboratory
Department of Biotechnology
University of North
Bengal, Siliguri-734013

A. Kumar
School of Biotechnology
Banaras Hindu University
Varanasi-221005

A. Saha
Molecular Plant Pathology and Fungal
Biotechnology Laboratory
Department of Botany
University of North Bengal
Siliguri-734013

A. Sen
NBU Bioinformatics Facility
Department of Botany
University of North Bengal
Siliguri-734013

A.J. Lloyd
Institute of Environmental and
Rural Science, Aberystwyth University
Edward Lloyd Building
Aberystwyth, UK, SY23 3DA

A.K. Bothra
Bioinformatics Chemoinformatics
Laboratory, Department of Chemistry,
Raiganj University College
Raiganj, India

A.K. Rai
School of Biotechnology
Banaras Hindu University
Varanasi-221005

A.K. Roy
P.G. Dept. of Biotechnology
T.M.Bhagalpur University,
Bhagalpur-812007

A.M. Chavan
Seed Pathology and Fungal
Biotechnology Laboratory
Dept. of Botany, Dr. Babasaheb
Ambedkar Marathwada University
Aurangabad-431004 (M.S.)

A.P. Chakraborty
Immuno-Phytopathology Laboratory
Department of Botany
University of North Bengal,
Siliguri-734013, West Bengal

B.N. Chakraborty
Immuno-Phytopathology Laboratory,
Department of Botany,
University of North Bengal
Siliguri-734013

B. Pradhan
Plant Biochemistry Laboratory
Department of Botany
University of North Bengal
Siliguri-734013, West Bengal

B. Saha
Molecular Plant Pathology and
Fungal Biotechnology Laboratory
Department of Botany
University of North Bengal
Siliguri-734013

C. Srinivas
Department of Microbiology and Biotechnology, Bangalore University
Jnana Bharathi Campus
Bangalore-560056

D. Nirmala Devi
Department of Microbiology & Biotechnology, Bangalore University,
Jnana Bharathi Campus
Bangalore-560 056

D. Pradhan
Plant Biochemistry Laboratory
Department of Botany
University of North Bengal
Siliguri-734013, West Bengal

D. Saha
Plant Biotechnology laboratory
Department of Biotechnology
University of North Bengal
Siliguri-734013, West Bengal

D.P. Gadgile
Seed Pathology and Fungal Biotechnology Laboratory
Dept. of Botany, Dr. Babasaheb Ambedkar Marathwada University
Aurangabad-431004 (M.S.)

H.D. Mohankumar
Horticulture Research Station
Devihosur, Haveri-581110

K. Jarial
Department of Mycology and Plant Pathology, Dr. Y.S. Parmar
University of Horticulture and Forestry
Nauni, Solan-173230

K. Rai
Immuno-Phytopathology Laboratory
Department of Botany
University of North Bengal
Siliguri-734013, West Bengal

K. Sunar
Immuno-Phytopathology Laboratory
Department of Botany
University of North Bengal
Siliguri-734013, West Bengal

L.A.J. Mur
Institute of Environmental and Rural Science, Aberystwyth University
Edward Lloyd Building
Aberystwyth, UK SY23 3DA

M. Sharma
Department of Mycology and Plant Pathology
Dr. Y.S. Parmar University of Horticulture and Forestry
Nauni, Solan-173230

M. Shivaprasad
Horticulture Research Station
Devihosur, Haveri-581110

M.B. Tyagi
Department of Botany
Centre of Advanced Study in Botany
Banaras Hindu University
Varanasi-221005

M.H. Tatagar
Horticulture Research Station
Devihosur, Haveri-581110

M.K. Sateesh
Molecular Diagnostic Laboratory
Department of Microbiology and Biotechnology, Bangalore University
Bangalore-560056

N.L. Mandal
P.G. Department of Biotechnology
T.M. Bhagalpur University
Bhagalpur-812007

P. More
Horticulture Research Station
Devihosur, Haveri-581110

P. Ray
Immuno-Phytopathology Laboratory
Department of Botany
University of North Bengal
Siliguri-734013, West Bengal

P.L. Dey
Immuno-Phytopathology Laboratory
Department of Botany
University of North Bengal
Siliguri-734013, West Bengal

R. Das Biswas
Immuno-Phytopathology Laboratory
Department of Botany
University of North Bengal
Siliguri-734013, West Bengal

R. Lama
Plant Biochemistry Laboratory
Department of Botany
University of North Bengal
Siliguri-734013, West Bengal

R. Som
Immuno-Phytopathology Laboratory
Department of Botany
University of North Bengal
Siliguri-734013, West Bengal

R.B. Kakde
Seed Pathology and Fungal
Biotechnology Laboratory
Department of Botany
Dr. Babasaheb Ambedkar
Marathwada University
Aurangabad-431004 (M.S.)

R.K. Mesta
Horticulture Research Station
Devihosur, Haveri-581110

R.P. Sinha
Centre of Advanced Study in Botany
Banaras Hindu University
Varanasi-221005

S. Allay
Immuno-Phytopathology Laboratory
Department of Botany
University of North Bengal
Siliguri-734013, West Bengal

S. Khati
Immuno-Phytopathology Laboratory
Department of Botany
University of North Bengal
Siliguri-734013, West Bengal

S. Rana
Immuno-Phytopathology Laboratory
Department of Botany
University of North Bengal
Siliguri-734013, West Bengal

S. Rana
Department of Mycology and
Plant Pathology
Dr. Y.S. Parmar University of
Horticulture and Forestry
Nauni, Solan-173230

S. Roy
Plant Biochemistry Laboratory
Department of Botany
University of North Bengal
Siliguri-734013, West Bengal

S. Sur
NBU Bioinformatics Facility
Department of Botany
University of North Bengal
Siliguri-734013, West Bengal

S. Thakur
NBU Bioinformatics Facility
Department of Botany
University of North Bengal
Siliguri-734013, West Bengal

S. K. Gupta
Department of Mycology and
Plant Pathology
Dr. Y.S. Parmar University of
Horticulture and Forestry
Nauni, Solan-173230

S. K. Shahi
School of Biotechnology
Banaras Hindu University
Varanasi-221005

S.S. Mohammad
Molecular Diagnostic Laboratory
Department of Microbiology and
Biotechnology, Bangalore University
Bangalore-560056

T.R. Lakshmeesha
Molecular Diagnostic Laboratory
Department of Microbiology and
Biotechnology, Bangalore University
Bangalore-560056

U. Chakraborty
Plant Biochemistry Laboratory
Department of Botany
University of North Bengal
Siliguri-734013, West Bengal

U. De
Immuno-Phytopathology Laboratory
Department of Botany
University of North Bengal
Siliguri-734013, West Bengal

V.S. Thakur
Regional Horticultural Research
Station, Dr. Y.S. Parmar University of
Horticulture and Forestry
Mashobra, Shimla-171007

□□□

CONTENTS

Chapter 1

THE POTENTIAL OF CHEMICAL FINGERPRINTING FOR THE IMPROVEMENT OF TEA (*CAMELLIA SINENSIS*) CULTIVATION IN INDIA

L.A.J. MUR[1], A.J. LLOYD[1] AND B.N. CHAKRABORTY[2]

[1] Institute of Environmental and Rural Science, Aberystwyth University, Edward Lloyd Building, Aberystwyth, UK, SY23 3DA

[2] Immuno-Phytopathology Laboratory, Department of Botany, University of North Bengal, Department of Botany, Siliguri-734013, India

Introduction

Indian agriculture faces considerable challenges to increase yields against a background of elevated environmental stress. In 2005, agriculture accounted for 18.6% of the GDP in 2005, accounted for ~ 9% of all exports and employed 60% of the country's population. However, the Intergovernmental Panel on Climate Change (IPCC) suggest that the global mean temperature may increase between 1.4 and 5.8°C by 2100 and, due to its geographical position, this will impact particularly severely on India (Sathaye, 2006). In northern India, temperature rises would severely affect major tea and rice producing areas making droughts such as observed in 2002-03 much more frequent or, paradoxically, could increase rainfall in some areas. These scenarios are likely to increase both abiotic and biotic stress. For example, conditions of increased temperature and humidity would promote the pathogen and pest "blooms" which would increase disease pressure on crop plants (Evans *et al.*, 2008).

A number of strategies could be followed to counter increase crop stress. Most often new crop varieties are bred which are able to tolerate either multiple stresses or the most pertinent stress. However, this is a slow and costly process and may be difficult to achieve. The derivation of transgenic lines are an attractive option as they may comparatively rapidity drived (Bhatnagar-Mathur *et al.*, 2008) and have proven worth in an Indian context (Subramanian and Qaim, 2010), but this strategy has proven to be controversial to some stakeholders (Yamaguchi, 2007). Another alternative is the use of plant-growth-promoting rhizobacteria (PGPR), which has been particularly successful in India as a low cost and effective method to improve crop stress tolerance (Ahmad *et al.*, 2008; Basnet *et al.*, 2009; Chakraborty *et al.*, 2009; Karthiba *et al.*, 2010; Loganathan *et al.*, 2010; Siddiqui and Futai, 2009). Further besides increasing immunity, plant productivity is also improved (Lucy *et al.*, 2004; Lugtenberg and Kamilova, 2009; Yang *et al.*, 2009).

In each of these strategies, there remains a need to improve our understanding of the responses of plants to stress, particularly within agriculturally relevant contexts. As recently highlighted (Vinocur and Altman, 2005), our knowledge of stress-associated metabolism is very limited so this needs to extensively profiled in order to be successful in developing stress-tolerant crop plants. In the case of PGPR, this is often undertaken from empirical observations and the exploitation of these bacteria could be improved if the mechanisms through which these strains function were better understood.

Metabolomics is an approach where significant proportions of cell, organ or whole organism metabolite pools are sampled and differences related to the biological phenomena under investigation (Dixon *et al.*, 2006). Metabolite detection involves either Nuclear Magnetic Resonance (NMR) or Mass Spectroscopy (MS) based approaches, whilst key discriminatory metabolites may be discerned through multivariate statistical techniques (Seger and Sturm, 2007). Metabolomics could be of great assistance to Indian agriculture by identifying metabolites associated resistance which could serve as chemical "biomarkers" within breeding programmes or be used to inform PGPR-based strategies. These possibilities not withstanding, many metabolomic approaches demand the use of considerable and costly resources. However, metabolic fingerprinting techniques provide a cheap first-round screen from which chemical signatures of samples are obtained and compared. Fourier Transform-Infra Red

(FT-IR) spectroscopy (Fig. 1A) is a well-established fingerprinting method which can analyse almost any biological material or chemical compound. FT-IR is a reagent-less and non-destructive means of high-throughput analysis, where samples are homogenised in water and directly spotted onto a sample plate for high-through put screening (Fig. 1B). Minimal training is required for instrument operation and a single spectrum is obtained within seconds and thousands of spectra may be obtained within a day (Goodacre *et al.*, 2004).

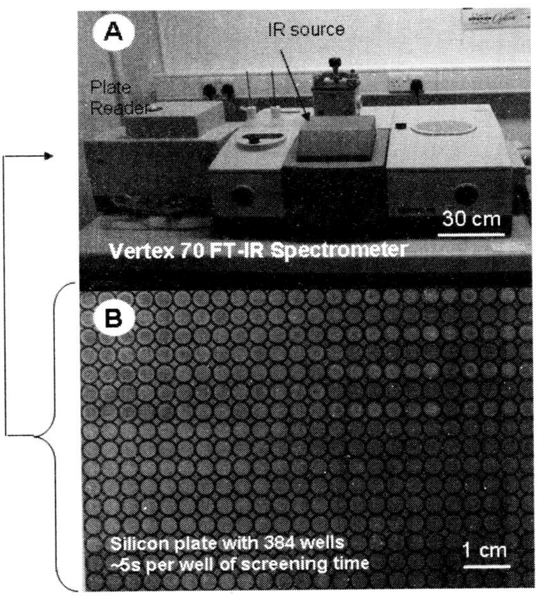

Figure 1: FT-IR Spectrometry. **(A)** Illustrated is the Bruker Vertex 70 FT-IR Spectrometer used in this study. The infrared (IR) source and insertion stage for the silica plates are arrowed. Bar = 30 cm **(B)** A 384 well silica plate. Bar = 1 cm.

In this paper, we describe the application of FT-IR fingerprinting to detect metabolic changes associated with Tea (*Camellia sinensis*) under biotic stress. Tea blister blight (causal agent, *Exobasidium vexans*), Tea brown blight (causal agent, *Glomerella cingulata*) and damage caused by the Tea mosquito, (*Helopeltis theivora*), represent major sources of loss in tea leaf quality and yields. We show that FT-IR fingerprinting can detect metabolite changes within samples of Indian field grown tea samples. These data will allow the detection of biomarkers associated with biotic stress and will serve as a basis for large scale metabolomic studies where the actual biochemical changes will be described.

Materials and Methods

Tea Sampling : Following infection of healthy young leaves of tea (Fig. 2A) growing within Margaret's Hope Tea Gardens (North Kurseong, Darjeeling, India, 26.95059N, 88.279537E; Altitude 950-1830m) with *Exobasidium vexans*, symptoms initially considered of a reddening of the infection site (Fig. 2B; designated stage 1), followed by formation of characteristic blister (Fig. 2C; designated stage 2) which subsequently leads to large necrotic patches in the leaves (Fig. 2D; designated stage 3). By contrast, infection with tea brown blight fungus (*Glomerella cingulata*) leads directly to the formation of large necrotic patches (Fig. 2D). These were sampled from Margaret Hope Tea Gardens.

Feeding by the tea mosquito (*Helopeltis theivora*) (Fig. 2F) leads to the formation of piecing holes which are surrounded by areas of extensive reddening (Fig. 2G). Samples of young leaves of healthy and attacked leaves were taken from tea grown in the germplasm bank (North Bengal University). Twelve genotypes were examined representing North Indian Tocklai (TV-9, TV-17, TV-22, TV-23, TV-26, TV-29, and TV-30) and South Indian UPASI tea varieties (UP-2, UP-3, UP-9, UP-26, and BSS-2).

Ten Samples each of young healthy leaves (Fig. 2A) and from blister exhibiting leaves (Fig. 2B, C, D), from *G. cingulata* infected leaves (Fig. 2E) and *Helopeltis theivora* (Fig. 2F) attacked leaves were collected. The ten leaves from each stress class and healthy leaves at equivalent stages of development were separately pooled. This sampling was repeated 10 times so that there were 10 replicates from each stress, each representing a pooling of ten individual samples.

Sample Preparation: Each sample was grown to a fine powder in liquid nitrogen. The metabolites were extracted using a chloroform: methanol: water (22:56:22) mix which allowed the isolation of both polar and nonpolar metabolites. The slurry was centrifuged at 3600 x *g* to remove any particulate matter and the supernatants collected. The volatile components of the extraction mix were evaporated over a water bath at 45°C for 4-6 h as required. The aqueous solvent was removed by lyophilisation. The dried samples were then dispatched to Aberystwyth University in the UK. On receipt the samples were resuspended in methanol: water (70:30) and stored at -80 °C until assessed using FT-IR spectroscopy.

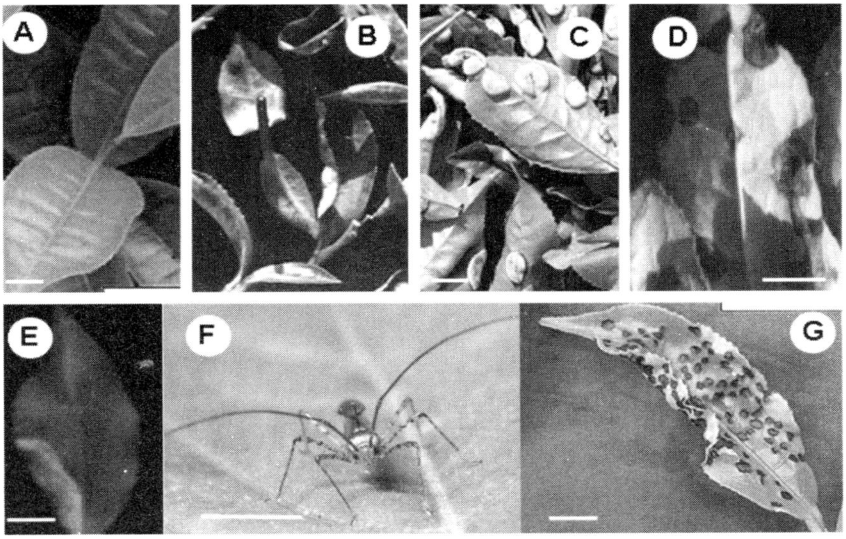

Figure 2: Biotic stress on tea (*Camellia sinensis*) leaves.

(A) healthy tea leaves and tea blister (*Exobasidium vexans*) infected leaves at designated infection stages **(B)** 1, **(C)** 2 and **(D)** 3. **(E)** A tea leaf infected with brown blight fungus (*Glomereliia cingulata*). **(F)** a tea mosquito (*Helopeltis theivora*) and **(G)** tea leaf damage caused by *H. theivora*. Bar = 1 cm.

***FT-IR spectroscopy*:** This was undertaken in accordance with the manufacturer's instructions (Bruker BioSciences Corp, Billerica MA USA). The FT-IR mid-infrared spectrometer instrument employed comprised: an IR radiation source; a Michelson interferometer; a diffuse reflectance absorbance scanning accessory equipped with a mercury-cadmium telluride (MCT) detector; and a sampling compartment fitted with a horizontal attenuated total reflectance (HATR) sampling accessory equipped with a deuterated triglycine sulphate (DTGS) detector. Windows XP computer was used for instrument control, Fourier transformation and data acquisition, using OPUS version 2.1 (provided by Bruker). For reflectance measurements of opaque samples, the MCT detector was used which was cooled with liquid N_2 prior to use and calibrated using a HTS Ratio Test with a calibration plate.

Spectra were created with a resolution of 4 cm^{-1} and a sample scan time of 64 scans permitting collection of 1763 data points over a wavenumber range of 4000 cm^{-1} to 600 cm^{-1}. The IR signal was converted to a digitised format by an analogue to digital converter (ADC) before being transferred to the computer for fast Fourier transformation yielding a wavenumber-dependant IR spectrum. The sample absorbance spectrum was calculated from the ratio of I_S/I_R, where I_S was the intensity of the IR beam after it has been absorbed by the sample and I_R was the intensity of the IR beam from the reference plate. The absorbance spectrum was calculated as $-\log_{10}(I_S/I_R)$.

Multivariate analyses: The FT-IR spectra that were obtained were exported in ASCII format from the instrument manufacturer's software and imported into MATLAB version 6.5 (The Math Works Inc., Natwick, MA, US). Each spectra were "smoothed" over 5 points using the Savitzky-Golay algorithm. The data were first analysed using principal components analysis (PCA). Principal components (PC) are linear combinations of original variables which can be used to explain the variation with a data matrix- in this case multiple FT-IR spectra. PC1 explains most of the variation within samples whilst PC2...*etc*, progressively explain less of the variation. PCA is known as an "unsupervised" method since it describes natural variation within a matrix irrespective of the experimental structure. In this study, PCA was followed by discriminant function analysis (DFA) which is a "supervised" projection method (Goodacre *et al.*, 2004). With PC-DFA the algorithm contains a description of the experimental class/data structure and variation between classes is specifically targeted. All calculations were performed in MATLAB (The Math Works Inc.).

Results and Discussion

Initial experiments focused on the samples extracted from young healthy tea leaves (Fig. 2A) and tea blister exhibiting leaves at each designated stage of infection (Fig. 2B, C and D). FT-IR was employed to assess whether the metabolite fingerprints differed between the different stages of infection. FT-IR is a well-established method through which a snap-shot of sample biochemistry may be obtained. FT-IR can be used in the non-destructive, low cost and high-throughput analysis of almost any biological material or chemical compound. Thousands of spectra can be obtained within a single day, requiring minimum sample preparation and background training. FT-IR is a physico-chemical method where a sample is interrogated with IR light and bonds and functional groups within the molecules absorb the IR at specific wavelengths, resonating in a predetermined way: stretching or bending (Griffiths, 1983). FT-IR spectroscopy of blister infected samples yielded information-rich spectra (Fig. 3A) but these provide to be impossible to discriminate using the naked eye and required multivariate statistical analysis to interpret.

Multivariate analyses frequently employed in metabolomic studies are divided into non-supervised and supervised methods (Goodacre *et al.*, 2004). Non-supervised techniques cluster the metabolite data into groups due to natural variation within the data. Principal component analysis (PCA) is non-supervised method that is commonly applied to multivariate data to reduce the dimensionality with minimal information loss. PCA enables the visualisation of the natural variability within data in the form of 2D or 3D plots of different combinations of underlying variables, known as principal components (PC; (Johnson *et al.*, 2003). The method has the advantage that data does not have to be normaily distributed. PCA is also often used to reduce the dimensionality of the data prior to further approaches being followed (Johnson *et al.*, 2003). These additional approaches include supervised methods, which require *a priori* knowledge of the class structure of the data. Discriminant Function Analysis (DFA) or Canonical Variate Analysis (CVA), Genetic Algorithms (GAs) and Partial Least Squares (PLS) regression, are all such methods. DFA is commonly applied to reduced data from PCA and aims to maximise the variance between classes, while minimising variance within classes (Goodacre *et al.*, 2004). If clustering is observed across a particular Discriminant Function (DF), then loading vectors can be examined; and highly loaded variables can be inferred to be important for the observed differences. These key components can be identified and modulated metabolites can be related back to the question being tested.

When PCA was applied to the tea blister data it was not possible to discern any clustering that could be related to the stages of infection (data not shown). This indicated that the greatest variation in the data was not associated with chemical changes linked to infection with *E. vexans* and was perhaps to be expected as the samples were gathered from a tea garden where in environmental conditions would be highly variable. In contrast, analysis of the data using DFA successfully identified variation within the datasets that were associated with infection stage (Fig. 3B). DF1 described some variation associated with infection stage whilst DF2 described changes linked to infection; irrespective of stage.

As clustering is observed across a particular DF, it is appropriate to examine the associated "loading vectors". Put crudely, each experimental variable is given an "eigenvalue", representing the relative contribution ("loadings") each variable has made to the derived model. These loadings may be plotted as a visual representation of the contribution of each experimental variable to the model. In Fig. 3C, the relative loadings for each experimental variable (in this case wavenumbers) used in DF2 to separate infected from healthy material are shown. Wavenumbers making major contributions to DF2 in the model shown in Fig. 3B are indicated by peaks and troughs about the 0 line (where variables [wavenumbers] make no contribution to the model).

Although FT-IR cannot identify individual metabolite the mid-IR region (4000-600 cm^{-1}) used here still allows the collection of extremely rich and useful biochemical information. In particular, in four major regions strong absorption can be directly related to a class of compounds (Fig. 3C; Table 1). These regions are broadly termed as the phenolic region (~3600-3200 cm^{-1}) fatty acid region (~2956 to 2850 cm^{-1}), the amide regions (~1745 to 1655 cm^{-1}) and the polysaccharide region (~1085 to 1052 cm^{-1}) (Nakanishi, 1962). It should be noted that FT-IR spectroscopy also includes the measurement of proteins (amide regions) and nucleic acids (~1240-1220 cm^{-1}), which are not usually considered in the metabolomic analyses. Examining the loadings associated with these distinctive IR regions, changes in the amide regions appear to be particularly prominent (Fig. 3C). Next, the IR absorption values for these distinctive regions (phenolic, fatty acid, amide and pyranose sugars; Fig. 3C) were extracted from the spectra for each sample. These subsets of data were then investigated to see it these would be sufficient to differentiate between healthy and infected tea leaves. DFA of the subset demonstrated that changes in these metabolite groups could explain differences between

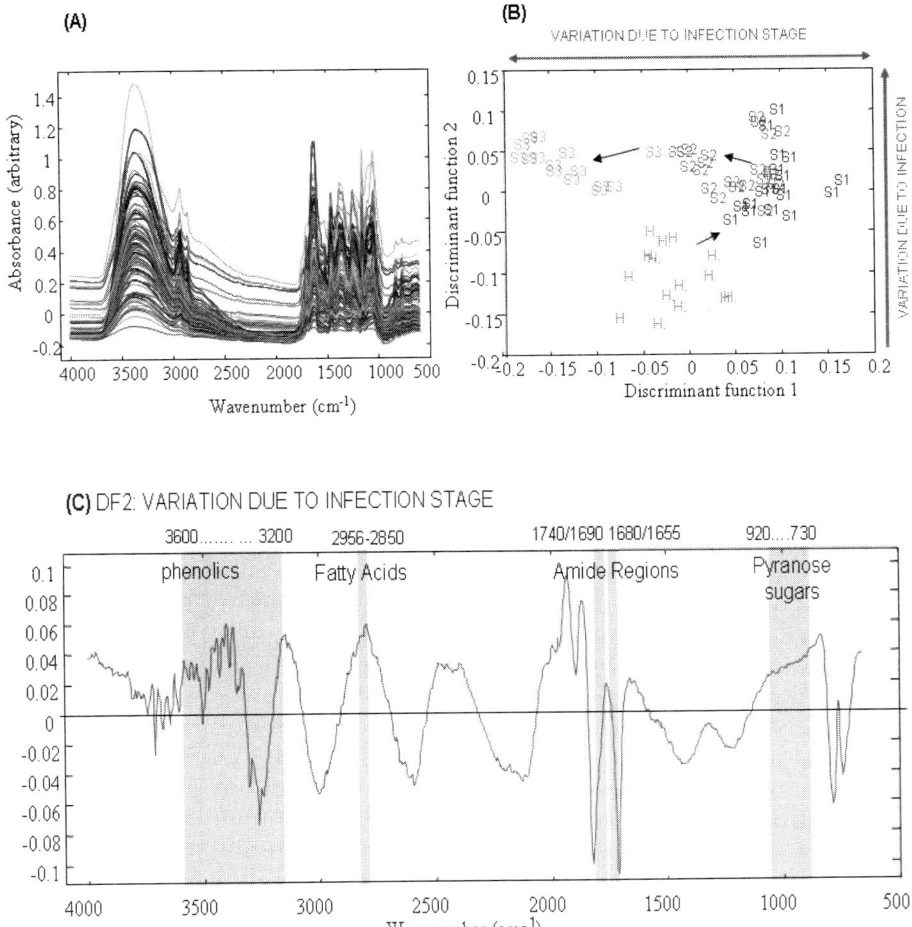

Figure 3: FT-IR analysis of Tea Blister (*Exobasidium vexans*) infected tea leaves. (A) FT-IR spectra derived from healthy and tea blister infected tea leaves at designated infection stages (see Materials and Methods). Each coloured line represents a spectrum from a single spectrum and does not reflect any biological class e.g. healthy or infected. (B) Discriminant Function Analysis of 10 samples from each biological class; H = Healthy; S1 = stage 1; S2 = stage 2 and S3 = stage 3. The model is based on 10 PC (principal components) which encompassed 99.23% of the total variance amongst the samples. Discriminant Function axes discriminating metabolite differences at different stages of infection (Discriminant Function 1) and changes associated with infection (Discriminant Function 2) are arrowed and labelled. (C) A loadings plot for DF2 showing the extent to which experimental variables (here wavenumbers) were used to derive the DFA model shown in (B). The shaded areas show wavenumber ranges associated with specific metabolite classes.

uninfected and infected tea leaves (Fig. 4A). The mean absorbance for each metabolite group for samples from healthy and each infection stage were derived. These displayed using a heat map (Fig. 4 B) where more intensive shades of red indicate comparatively higher absorbencies and deeper shade of blue comparatively lower absorbencies. These values were also related using a dendrogram derived using a hierarchical cluster analysis (HCA). These analyses showed that the sugar, amide and fatty acid content decreased within samples on infection whilst phenolics increased (Fig. 4B).

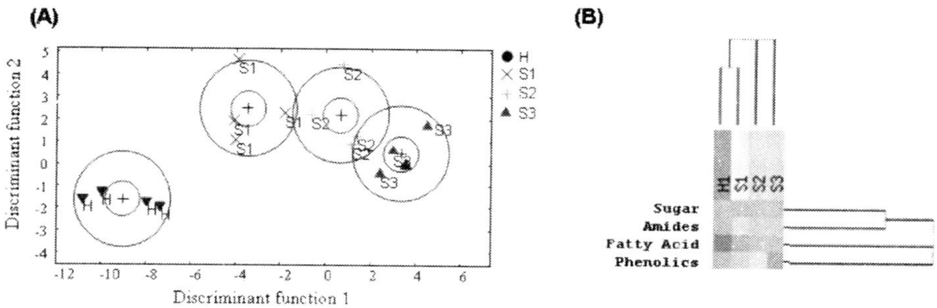

Figure 4: FT-IR analysis of Tea Blister (*Exobasidium vexans*) infected tea leaves based on IR absorbencies linked to specific metabolite classes.

Absorbences of certain regions of the FT-IR spectrum are linked to specific metabolite classes; pyranose sugars (1085-1055 cm^{-1}), amide regions (1680-1655/ 1740-1690 cm^{-1}), fatty acid (2956-2850 cm^{-1}) and phenolics (3600-3200 cm^{-1}). For each sample the IR absorbencies within these wavenumber ranges were extracted and compared. (A) Discriminant Function Analysis (DFA) of IR absorbencies of specific metabolite classes for samples from healthy (H) and tea blister (*Exobasidium vexans*) infected tea leaves at designated stage 1 (S1), stage 2 (S2) and stage 3 (S3). The model is based on 10 PC (principal components) which explains 99.19% of the total variance amongst the samples. The larger and smaller circles encompass 95% and 90% confidence internals with the mean value for each biological class shown by a cross. (B) Average IR spectral absorbencies were observed for each metabolite class and for each sample class (H, S1, S2 and S3). These were analysed as a hierarchical cluster analysis (HCA) with

Table 1 : Characteristic regions within Fourier Transform-Infra Red (FT-IR) spectra and absorbencies for key functional groups.

Region	Wavenumber (cm^{-1})	Band assignment
Fatty acid region	2956	CH_3 asym. stretch
	2920	CH_2 asym. stretch
	2870	CH_3 sym. stretch
	2850	CH_2 sym. stretch
Amide region	1745-1735	>C=O stretch, ester, fatty acids
	1705	>C=O stretch, ester, carboxylic groups
	1652-1648	Amide I, (C=O)
	1550-1548	Amide II, N-H, C-N and proteins
	1460-1454	C-H bend, from CH_2
Mixed region	1400-1398	C-O bend from carboxylate ions
	1303	Amide III (C-N)
	1240	P=O from phosphate
	1222	P=O
	1114	C-O-P, P-O-P
Polysaccharide region	1085	Ring vibrations
	1052	C-O, C-O-C from polysaccharides

the relative distances between the average values, between metabolite and biological classes as indicated using horizontal and vertical dendrograms respectively. The average values are indicated relatively using arbitrary scales using in heat map. In the heat map progressively increased values is shown by an intensification of red whilst progressively reduced levels by intensification in blue colour.

Following these successes, we next investigated other forms of biotic stress imposed on tea. IR spectra of samples extracted from tea leaves exhibiting attack by tea brown fungus (*G. cingulata*) were easily distinguished from healthy leaves (Fig. 5A). Similarly, DFA analyses of FT-IR spectra of samples from leaves attacked by the tea mosquito (*H. theivora*) were very different from healthy leaves (Fig. 5B). In Fig. 5B, the DFA model was based on two groupings – samples from infected and healthy tissue. However, it was possible to impose further

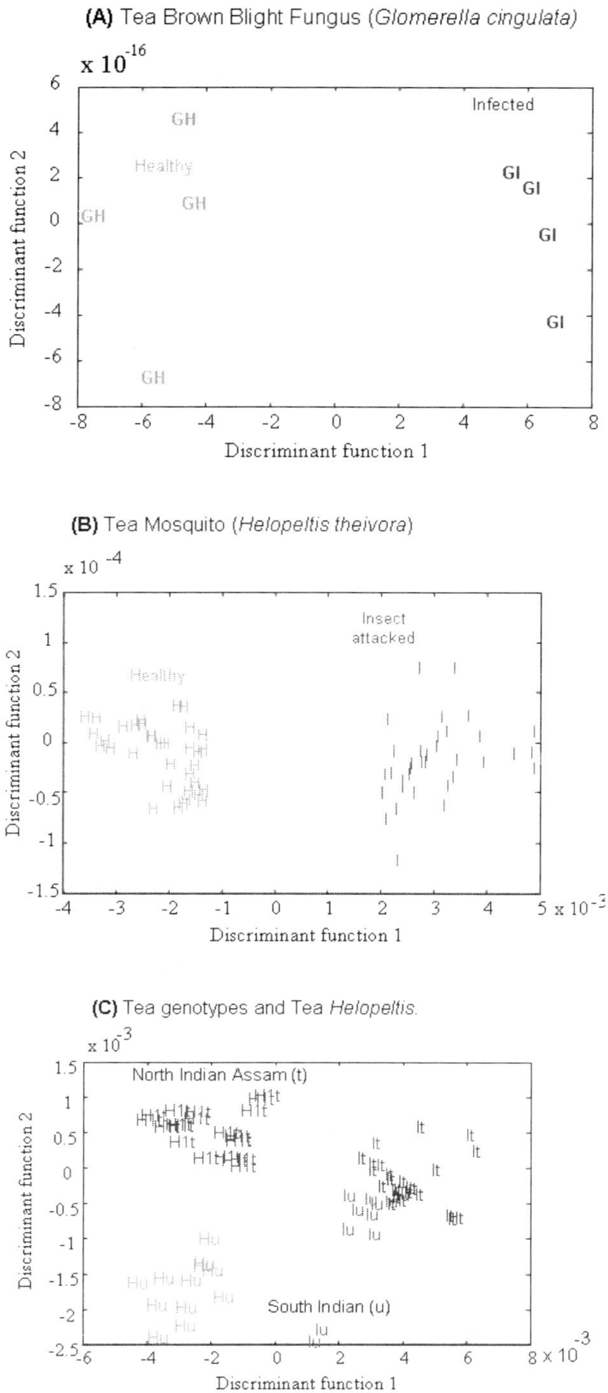

Figure 5: FT-IR analysis of Tea Brown Blight fungus (*Glomerella cingulata*) infected tea leaves and Tea Mosquito (*Helopeltis theivora*) attacked tea leaves. **(A)** Discriminant Function Analysis (DFA) of healthy tea leaves (GH) and tea brown blight infected (*Glomerella cingulata*) (GI) leaves. The model is based on 10 PC (principal components) which encompassed 99.44% of the total variance amongst the samples. **(B)** DFA of healthy tea leaves (H) and Tea Mosquito (*Helopeltis theivora*) attacked tea leaves (I). The model is based on 10 PCs which encompassed 99.06% of the total variance amongst the samples. **(C)** DFA of tea leaves from North Indian Assam (indicated by a "t" suffix) and South Indian (indicated by a "u" suffix) which were either healthy ("H" prefix) or attached by tea mosquito ("I" prefix) leaves. The model is based on 10 PCs which encompassed 99.38% of the total variance amongst the samples.

groupings by attempting to differentiate the responses of North Indian Tocklai and South Indian UPASI varieties of tea to tea mosquito (Fig. 5C). This revealed that FT-IR spectroscopy could distinguish between healthy leaves from North Indian and South Indian varieties but not in the responses of these varieties to tea mosquito. This suggests that the common responses of these tea varieties to tea mosquito effectively mask background varietal differences in metabolite profile.

These preliminary studies has demonstrated that metabolomic analyses can be applied to reveal key chemical changes occurring in Indian tea gardens, rather than just carefully regulated environmental conditions. They have immediate practical application in defining IR biomarkers which can be associated with either disease progression or resistance. As such is would be of great interest to combine these studies with tea cultivars where PGPR bacteria have applied to increase resistance to biotic stress. The samples are currently being analysed by MS so that those key metabolite differences can be described.

Acknowledgements

BN Chakraborty acknowledges the financial support of the Royal Society of Chemistry (East India Section) in developing this project.

References

Ahmad, F., Ahmad, I. and Khan, M. S. (2008) Screening of free-living rhizospheric bacteria for their multiple plant growth promoting activities. *Microbiological Research* 163: 173-181.

Basnet, M., Chakraborty, B.N. and Chakraborty, U. (2009) Exploitation of Tea Rhizosphere Microorganism for Improvement of Plant Health Status. *Journal of Mycology and Plant Pathology* 39: 1-13.

Bhatnagar-Mathur, P., Vadez, V. and Sharma, K. K. (2008) Transgenic approaches for abiotic stress tolerance in plants: retrospect and prospects. *Plant Cell Reports* 27: 411-424.

Chakraborty, U., Chakraborty, B. N., Basnet, M. and Chakraborty, A. P. (2009) Evaluation of *Ochrobactrum anthropi* TRS-2 and its talc based formulation for enhancement of growth of tea plants and management of brown root rot disease. *Journal of Applied Microbiology* 107: 625-634.

Dixon, R. A., Gang, D. R., Charlton, A. J., Fiehn, O., Kuiper, H. A., Reynolds, T. L., Tjeerdema, R. S., Jeffery, E. H., German, J. B., Ridley, W. P. (2006) Perspective - Applications of metabolomics in agriculture. *Journal of Agricultural and Food Chemistry* 54: 8984-8994.

Evans, N., Baierl, A., Semenov, M. A., Gladders, P. and Fitt, B. D. L. (2008) Range and severity of a plant disease increased by global warming. *Journal of the Royal Society Interface* 5: 525-531.

Goodacre, R., Vaidyanathan, S., Dunn, W. B., Harrigan, G. G. and Kell, D. B. (2004) Metabolomics by numbers: acquiring and understanding global metabolite data. *Trends in Biotechnology* 22: 245-252.

Griffiths, P. R. (1983) Fourier-Transform Infrared Spectrometry. *Science* 222: 297-302.

Johnson, H. E., Broadhurst, D., Goodacre, R. and Smith, A. R. (2003) Metabolic fingerprinting of salt-stressed tomatoes. *Phytochemistry* 62: 919-928.

Karthiba, L., Saveetha, K., Suresh, S., Raguchander, T., Saravanakumar, D. and Samiyappan, R. (2010) PGPR and entomopathogenic fungus bioformulation for the synchronous management of leaffolder pest and sheath blight disease of rice. *Pest Management Science* 66: 555-564.

Loganathan, M., Sible, G. V., Maruthasalam, S., Saravanakumar, D., Raguchander, T., Sivakumar, M. and Samiyappan, R. (2010) Trichoderma and chitin mixture based bioformulation for the management of head rot (Sclerotinia sclerotiorum (Lib.) deBary)-root-knot (Meloidogyne incognita Kofoid and White; Chitwood) complex diseases of cabbage. *Archives of Phytopathology and Plant Protection* 43.

Lucy, M., Reed, E. and Glick, B. R. (2004) Applications of free living plant growth-promoting rhizobacteria. *Antonie Van Leeuwenhoek International Journal of General and Molecular Microbiology* 86: 1-25.

Lugtenberg, B. and Kamilova, F. (2009) Plant-Growth-Promoting Rhizobacteria. *Annual Review of Microbiology* 63: 541-556.

Nakanishi, K. (1962) Infrared Absorption Spectroscopy Practical.: Holden Day Inc. (San Francisco) and Nankodo Company Ltd. (Tokyo).

Sathaye, J., Shukla, P.R. and Ravindranath, N.H. (2006) Climate change, sustainable development and India: Global and national concerns. *Current Science* 90: 314-325.

Seger, C. and Sturm, S. (2007) Analytical aspects of plant metabolite profiling platforms: Current standings and future aims. *Journal of Proteome Research* 6: 480-497.

Siddiqui, Z. A. and Futai, K. (2009) Biocontrol of Meloidogyne incognita on tomato using antagonistic fungi, plant-growth-promoting rhizobacteria and cattle manure. *Pest Management Science* 65: 943-948.

Subramanian, A. and Qaim, M. (2010) The Impact of Bt Cotton on Poor Households in Rural India. *Journal of Development Studies* 46: 295-311.

Vinocur, B. and Altman, A. (2005) Recent advances in engineering plant tolerance to abiotic stress: achievements and limitations. *Current Opinion in Biotechnology* 16: 123-132.

Yamaguchi, T. (2007) Controversy over genetically modified crops in India: discursive strategies and social identities of farmers. *Discourse Studies* 9: 87-107.

Yang, J., Kloepper, J. W. and Ryu, C. M. (2009) Rhizosphere bacteria help plants tolerate abiotic stress. *Trends in Plant Science* 14: 1-4.

□□□

Chapter 2

PRESENT SCENARIO OF MICROBIAL DIVERSITY OF COAL SPOILS AND PROSPECT FOR RECLAMATION OF COAL FIELD AREAS

A.K. ROY AND N. L. MANDAL

P.G. Dept. of Biotechnology, T.M.Bhagalpur University, Bhagalpur-812007

Introduction

Microorganisms represent the richest source of biological diversity inhabiting all types of ecosystem either of tropical, temperate or extreme, of which rhizosphere and soil microbes like bacteria, fungi, mycorrhizae, yeasts, and actinomycetes have great role in plant health as well as disease management. The health of soil is also determined by microbial flora due to their immense potential for nutrient recycling, mobilization as well as plant growth promotion by the release of growth hormone and siderophore. For the development of greenery or agriculture on sustainable basis it is essentially required to evaluate the role of microbes on plant growth without affecting the bioresources, environmental quality and economical losses. On account of variable microbial diversity and their efficacy scientists of different disciplines are always in search of beneficial microbes from various sources and also modifying them genetically which may be more specific for plant health. In view of these facts the present investigation has been aimed to isolate microbes from different coal mine areas viz., Lalmatia (ECL), Singrauli (NCL), Samleshwari (MCL) and evaluate their role in plant health.

Coal mining activity: Out of the total coal production, more than 50% is procured from open cast mining alone. Opencast excavation of coal deposits involves the removal of overlying soil and rock debris and their storage as overburden dumps (OBD). Generation of OBD during open cast operation is inevitable, which consists of stone pebbles, coal spoils along with nutritionally poor soil, water deficient and often lacking in the necessary organic matters essential for the establishment of micro and macro flora. However, in due course of time soil particles settle down, allowing natural vegetation to grow. Opencast mining operations have a number of irreversible impacts on the surrounding environment and ecosystems. Thus OBD represents an extreme physico-chemical profile of soil along with disturbed dynamics of micro flora obviously due to broken myriad of tight and tenuous links between vegetation and soil microorganisms. Mitigation measures commonly adopted in mined out areas are: compensatory afforestation /plantation, reclamation, rehabilitation and ecological restoration. These overburden dumps change the natural land topography, affect the drainage system and prevent natural succession of plant growth (Bradshaw and Chadwick 1980; Wali 1987) resulting in acute problems of soil erosion and environmental pollution (Singh *et al.* 1994; Singh *et al.* 1996). Therefore, self-sustainability and regenerative capacities of abandoned mining areas should be developed for proper functioning of the ecosystems through revegetation (Jha and Singh 1994).

Coal mining not only visibly disrupts the aesthetics of the landscape but it also changes soil components such as soil horizons and structure, soil microbe populations, and nutrient cycles that are crucial to sustaining a healthy ecosystem. Ecosystems function optimally as a process of recycling nutrients balance of growth rates and decomposition rates of plant and animal matter. Soil provides the foundation for this process, so its composition and density directly affect the future stability of the restored plant community.

Microbial diversity in coal spoils

Microorganisms exhibit extraordinary genetic and phenetic diversity on land, in fresh water or sea or as parasites or symbionts. Despite their importance, less than 5% of the world's microorganisms have been described. Of the 300,000 to one million species of prokaryotes on earth only 3,100 bacteria have been described in the Bergey's Manual of the Systematic Bacteriology. The omnipresence of micrograniṡms through the biosphere is a consequence of their ready dissemination by wind and water. Surface waters, the floors of oceans over the continental shelves and the top few inches of soil are teeming with microorganisms that are ready to decompose organic matter that may become available to them. The abundance of prokaryotes in open ocean, in soil and in oceanic and terrestrial subsurfaces is 1.2 x 1029, 2.6 x 1029, 3.5 1030 and 0.25-2.5 x 1030 respectively (Whitman *et al*, 1998). Among the deep-sea microflora approximately one third of the prokaryotes are Archaea (Karner *et al*, 2001). The estimated number of viruses in the world oceans is approximately 7.5 x 1029 while no total virus number are known from soil and subsurface (Weinbauer and Hofle, 1998). The top 6 inches of fertile soil contains more than 2 tons of fungal and bacterial biomass. It has been estimated that there might be up to 10,000 bacterial species per 100 g soil (Torsvik *et al*, 1990). A handful of soil contains different kinds of microbes; even a single soil particle represents ecological niches for different types (Nee, 2002; Steage and Zaft, 2002; Morin, 2000). Thus the prokaryotes have been described correctly as the unseen majority (Whitman *et al*, 1998) and this is more correct for all microbes.

At the early stages of an ecosystem development, soil acts as a critical controlling component. Without the natural processes of soil development, ecosystems would remain in a degraded condition. Mine spoil heaps are composed of coarse rocks due to the deep coal mining operations and associated coal processing. These spoils are not suitable for either plant or microbial growth because of low organic matter content, unfavourable pH, drought arising from coarse texture or oxygen deficiency due to compaction (Agrawal *et al.* 1993). The other limiting factors for revegetation of mine spoil may be salinity, acidity, poor water holding capacity, inadequate supply of plant nutrients and accelerated rate of erosion (Jha & Singh 1991).

Microorganisms are known to have potentials for the establishment of plants in the non-culturable lands. Since soil microflora have pivotal role in mobilizing soil nutrients and producing growth promoting substances either independently or in association, they are important in regeneration of mined spoils. Successful biological reclamation strategy depends on the selection of appropriate plant and microbial species, in addition to climate, physical and chemical properties of dump materials, topography and viability of surrounding vegeta-tion (Singh and Jha, 1992). Further due to variation in agro-climatic condition, vegetation and microbial diversity as well as their population are not alike throughout the country ; also the growth and survivability of microbes are varied from plant to plant, soil to soil and also to different OBD soil. Therefore, to develop vegetation in a particular zone or OBD it is essential to take up the native biological resources in and around the vicinity of coal field area for its reclamation.

During the reclamation of mine spoil, it is often necessary to establish and maintain a vegetative cover without the use of top soils or other bulky amendments (Rimmer 1982). Pederson *et al.* (1988) observed that the surface material from coal mining in Pennsylvania had a high density and low porosity resulting in low infiltration rates. In addition to soil physical characteristics, the nutrient status of overburden soil is also a major factor limiting plant growth. The cycling of nutrients regulates the sustainability of any plant community. Without cycling, nutrients are lost or immobilized and the plant community is not capable of regeneration. Berg & Barrau (1978) suggested that rapid re-establishment of the nitrogen cycle appears to be particularly important, but often difficult to achieve in mine spoil, as its geomorphic system is in disequilibrium due to the destruction between land form and processes resulting in acceleration of erosion rate (Dutta 1999). Mine spoils present very rigorous conditions for both plant and microbial growth because of low nutrient contents and coarse texture (Dutta & Agrawal 2000). Natural plant succession is also very slow on coal mine spoil land, however raising of plantation may accelerate this process leading to a self sustained ecosystem in a relatively short period of time (Singh & Singh 1999). The data of microbial diversity as presented in Table 1 clearly reveals changes in their population after plantation. It is also noticed that ageing of plantations favour increase in microbial diversity.

Table 1 : Microbial diversity in coal spoils

Parameter	Before plantation	After plantation 6 months	12 months	18 months
Mycorrhizal spores /10g dry soil	13	32.0	42.0	46.0
N_2-fixing bacteria (CFUx10^4)	1.45	1.68	1.97	2.14
Total bacterial count (CFUx10^4)	4.10	4.98	5.70	6.23

Microbial potentials of Soil: Changes in microbial diversity are a natural biological event but its study is essential for the development of biofertilizer package for different coal mine areas which has also been studied in ECL, NCL and MCL. Biofertilizers are the source of microbial inoculants, which have hopes for many countries both economically and environmentally. Therefore, in developing countries like India, biofertilizers can solve problems of high cost of fertilizers and thus can save the economy of the country (*Gupta et al., 2003*). Bio-fertilizers, in strict sense, are not fertilizers which directly give nutrition to crop plants. These are cultures of microorganisms like bacteria, fungi, packed in a carrier material. Thus, the critical input in Biofertilizers is the microorganisms. There are various types of biofertilizers viz, **(a) Nitrogen-fixing biofertilizers** (*Rhizobium, Azotobacter, Azospirillium,* Blue-green algae, *Azolla)* **(b) Phosphorus mobilising biofertilizers (i) Phosphate solubilizer** *(Bacillus, Pseudomonas, Aspergillus niger)* (ii) **Phosphate absorber**-VAM fungi - e.g., *Glomus, Gigaspora* **(c) Organic matter decomposer biofertilizers** (i) *Cellulolytic (Cellulomonus,- Trichoderma)* (ii) *Lignolytic (Arthrobacter, Agaricus)* which help the plants indirectly through better Nitrogen (N) fixation or improving the nutrient availability in the soil. In the present investigation different combination of biofertilizer as mentioned in Table-5 have been tried (T_1-T_9) for the reclamation of coal field areas. Out of 14 plants (Table-2) selected for this purpose 10 each for MCL and NCL and 11 for ECL were planted on OBDs of respective coal field areas by amending the OBD's soil with various combination of biofertilizer. T_6 combination promoted growth for all plants except *Dendrocalamus strictus* for which T_4 combination was found to promot plant growth to the highest level (Table-6).

Constraints for reclamation of coal field areas: In spite of physical constraints i.e., unfavourable porosity, aeration, water holding capacity and absence of structural aggregates in the dumps; chemical constraints i.e., presence of poor organic and inorganic plant nutrients in coal spoils and biological constraints i.e., the absence of beneficial microbes, the waste land patches developed by open cast mining is ably reclaimed through selected plant species and bio-inoculant package. The rejuvenation of mine spoil dump and mined land productivity and fertility through amendment of organic material, biofertilizers and endomycorrhizal fungi favour restoration of the degraded land ecosystem and also helps to create luxuriant forests in the country providing substantial carbon dioxide sink vital for biosphere integrity, built up of nutritive top soil and enhanced ground water recharge (Juwarkar, 2004). It is well established that the OBD suffers both with physical and chemical constraints which does not favour speedy vegetation covers over there but plantation through bio-inoculant package changes the constraint into conducive physical properties and nutrient potentials of the OBD soil. The analysis of soil in pre and post plantation condition gives a clear picture as shown in table 3& 4 that changes are supportive to plant growth too.

Table 2 : Selection of plants for successful reclamation programme

Sl. No.	Name of Plant Species	ECL	MCL	NCL
1	*Cassia siamia*	✓	✓	✓
2	*Dendrocalamus strictus*	✓	✓	✓
3	*Dalbergia sissoo*	✓	✓	✓
4	*Gliricidia sepium*	✓	✓	✓
5	*Sesbania grandiflora*	✓	✓	✓
6	*Jatropha curcas*	✓	✓	✓
7	*Acacia cracicarpa*	✓	×	×
8	*Cymbopogon citrates*	✓	✓	✓
9	*Stylo hamata*	✓	×	×
10	*Stylo scabra*	✓	×	×
11	*Pterocarpus santalinus*	✓	×	×
12	*Terminalia arjuna*	×	✓	✓
13	*Terminalia bellirica*	×	✓	✓
14	*Terminalia chebula*	×	✓	✓

Table 3 : Phyical profile of OBD Soil

Parameter	Before plantation	After plantation 6 months	12 months	18 months
Sand (%)	45.99	44.03	42.67	40.0
Silt (%)	46.45	47.83	48.75	49.47
Clay (%)	7.66	8.14	8.58	9.16
Bulk density (g/cc)	1.23	1.19	1.10	1.07
WHC (%)	16.70	17.46	18.25	19.63
Porosity (%)	53.69	54.86	56.27	58.12
pH	7.72	7.23	6.68	6.54

Table 4 : Chemical profile of OBD Soil

Parameter(ppm)	Before plantation	After plantation 6 months	12 months	18 months
N	0.0119	0.134	0.0176	0.0183
P	3.02	4.78	6.27	6.98
K	71.68	68.39	64.76	62.94
S	21.50	23.89	26.20	28.13
Ca	7.00	6.12	5.87	5.46
Na	6.8	7.9	9.7	9.8
OC (%)	1.03	1.98	3.12	3.36

Table 5 : Different combinations of bio-inoculant package

Treatment	Treatment combination
T1	OBD Soil (Control)
T2	OBD soil + Vermicompost (1:1)
T3	OBD Soil + Vermicompost (2:1)
T4	OBD Soil + VC + VAM (1:1:1)
T5	OBD soil + VC + VAM (2:1:1)
T6	OBD Soil + VC + VAM + N2FB (1:1:1:1)
T7	OBD Soil + VC + VAM + N2FB (2:1:1:1)
T8	OBD Soil + VAM (1:1)
T9	OBD soil + N2 FB (1:1)

Table 6 : Growth parameters of treated plants during polybag experiment with OBD soil

S.N.	Plant	Treatments	Plant height Avg. (in cm.)		
			30 DAP	60 DAP	90 DAP
1.	*Bauhinia purpurea*	T6	9.7	12.4	16.8
2.	*Cassia siamea*	T6	16.4	20.5	25.6
3.	*Cymbopogon citratus*	T6	16.8	28.5	46.7
4.	*Dalbergia sissoo*	T6	15.9	20.4	30.4
5.	*Dendrocalamus strictus*	T4	14.6	23.4	30.7
6.	*Delonix regia*	T6	7.3	9.4	12.9
7.	*Gliricidia sepium*	T6	17.2	23.7	38.6
8.	*Jatropha curcas*	T6	18.3	21.5	23.7
9.	*Pongamia pinnata*	T6	8.6	11.3	16.3
10.	*Pterocarpus santalinus*	T6	17.4	19.6	27.6
11.	*Stylo hamata*	T6	10.8	13.8	18.2
12.	*Sesbania grandiflora*	T6	22.5	25.7	32.4
13.	*Terminalia arjuna*	T6	18.8	24.3	27.9
14.	*Terminalia bellirica*	T6	19.8	21.7	24.2
15.	*Terminalia chebula*	T6	16.9	21.3	31.8

Prospect of reclamation in coal mines: The major goals of the revegetation are to reduce erosion, stabilize mine spoils and develop processes for suitable post mining land use. It also promotes retrieval of micro flora which play important role to enrich the soil profile of OBD. Changes in microbial diversity are a natural biological event but its study is essential for the development of biofertilizer package of different OBD. There are several factors which are responsible for the effective reclamation in a particular coal mine with a suitable biofertilizer which are as follows:

- Suitability of the species to the target crop
- Suitability of the strain: There are specific strains of *Rhizobium* for different leguminous species like Cowpea, Red gram, Soybean, Alfalfa etc. Biofertilizer of specific culture should be used for specific crop.
- Identification of strains as suited to the agro-eco system, particularly the soil pH and moisture conditions. Through

research, specific strains as suited to a particular soil and environmental conditions are usually identified and pure mother cultures are maintained in research labs for supply to the commercial manufacturers.

- The aseptic conditions of manufacturing, the cell count of living organism present in the carrier material, purity and level of contamination.
- The conditions of carrier material in which the culture is packed and the quality of the packing material, which determine the shelf life.
- The conditions in which the packed materials are stored, distributed and kept with the farmers before it is applied.
- Soil conditions particularly pH, organic matter content, moisture level and agronomic practices.

Conclusion

Reclamation must go beyond planting a new landscape by considering the land as an integrated system that functions above and below the ground. Project plans must include strategies that rebuild soil structure, stimulate soil microbial populations and re-establish nutrient cycles if this system is to reassemble. Researchers have demonstrated techniques that appear successful over periods of several years and have indicated that there is much more to learn about their long term effects. It is unclear how frequently or for what length of time follow-up is to be done on reclaimed sites to compare their function with undisturbed sites. Funding continues to be available to coal mine operators and states to reclaim land on active and abandoned coal mine sites. Expanding follow-up efforts, particularly on sites that were abandoned for many years, may reveal valuable information that will guide the success of future reclamation activities.

References

Agrawal, M., J. Singh, A.K. Jha and J.S. Singh. (1993). Coal-based environmental problems in a low rainfall tropical region. pp. 27-57. *In*: R.F. Keefer & K.S. Sajwan (eds.). *Trace Elements in Coal Combustion Residues*. Lewis Publishers, Boca Raton.

Berg, B. and E.M. Barrau. (1978). Management approaches to nitrogen deficiency in revegetation of sub alpine disturbances. Information Series. 28. pp. 174-181. *In*: S.T. Kenny (ed.) *Proceedings of High Altitude Revegetation Workshop Number 3:* Colorado State University, Fort Collins, Colorado.

Bradshaw, AD and Chadwick, MJ (1980). The Restoration of Land. Blackwell Scientific Publications, Oxford pp. 210.

Dutta, R.K. and M. Agrawal. (2000). Reclamation of mine spoils: a need for coal industry. pp. 239-250. *In*: Arvind Kumar & P.K. Goel (eds.) *Industry, Environment and Pollution*. Technoscience Publications, Jaipur, India.

Dutta, R.K. (1999). *Performance and Impact of Selected Exotic Plant Species on Coal Mine Spoil.* Ph.D. Thesis. Banaras Hindu University, Varanasi, India

Gupta, R.K., Kaushik, S., Sharma, P. and Jain, V.K. (2003) Biofertilizers: An eco-friendly alternative to chemical fertilizers. *Environmental Challenges of the 21st Century*, APH Publishing Corporation, New Delhi, pp 275-287.

Jha, A.K. and Singh J.S. (1991). Spoil characteristics and vegetation development of an age series of mine spoils in a dry tropical environment. *Vegetation* 97: 63-76.

Jha, A.K and Singh, J.S (1994). Restoration of mine spoils. Reports on Concepts and Strategies. Green Era in Indian Steel – A Technical Compendium. Steel Authority of India, India.

Juwarkar, Asha A., Jambulkar, H.P. and Singh, S.K. (2004) *Appropriate strategies for reclamation and revegetation of coal mine spoil dumps.* JIPHE published on the occasion of National seminar on Environmental Engineering with special Emphasis on Mining Environment, ISM, Dhanbad, 19-20 March.

Karner MB, Delong EF and Karl DM (2001) Archaeal dominance in the mesopelagic zone of the pacific ocean. *Nature*, 409: 507-510.

Morin PJ (2000) Biodiversity -ups & downs. *Nature*, 405: 463-464.

Nee S (2002) Thinking big in ecology. *Nature* 417: 229-230.

Pederson, T.A., A.S. Rogowski and R. Pennock. (1988). Physical characteristics of some mine spoils. *Soil Science Society American Journal* 44: 131-140.

Rimmer, D.L. (1982). Soil physical conditions on reclaimed spoil heaps. *Journal of Soil Science* 33: 567- 579.

Singh, A N and J.S. Singh. (1999). Biomass, net productivity and impact of bamboo plantation on soil redevelopment in a dry tropical region. *Forest Ecology and Management* 119: 195-207.

Singh, JS and Jha, AK (1992). Restoration of degraded land: an overview. In Restoration of Degraded Land: Concepts and Strategies (ed. J.S. Singh), pp. 1-7. Rastogi Publications, Meerut, India.

Singh, RS, Tewary, BK and Dhar, BB (1994). Effect of surface mining on plant biomass and productivity in a part of Dhanbad coalfield areas. In Second National Seminar on Minerals and Ecology (ed. S.P. Banerjee), pp. 103-109. Oxford & IBH Pub., New Delhi.

Singh, RS, Chaulya, SK, Tewary, BK and Dhar, BB (1996). Restoration of a coal-mine overburden dump – a case study. Coal International, March.

Steage HT and Zaft R (2002) Density and diversity. *Nature* 417: 698-699.

Torsvik V, Goksoyr J and Daae FL (1990) High diversity in DNA of soil bacteria. *Appl Environ Microbiol*, 56: 782-787.

Wali, MK (1987). The structure dynamics and rehabilitation of drastically disturbed ecosystems. In Perspectives in Envi-ronmental Management (ed. T.N. Khoshoo), pp. 163-183. Oxford Publications, New Delhi.

Weinbauer MG and Hofle MG (1998) Distribution and life strategies of two bacterial populations in a eutrophic lake. *Appl Environ Microbiol*, 64: 3776-3783.

Whitman W, Coleman D and Wiebe W (1998) Prokaryotes: The unseen majority. Proc Natl Aca Sci USA, 95: 6578-6583.

❑❑❑

Chapter 3

APPLE DISEASES AND THEIR MANAGEMENT

V. S. THAKUR

Regional Horticultural Research Station, Dr. Y. S. Parmar University of Horticulture and Forestry, Mashobra, Shimla –171 007 (HP) INDIA

Introduction

Apple (*Malus domestica* Borkh.) is a premier table fruit of the world and has been under cultivation since time immemorial. It is a temperate tree fruit grown extensively in temperate parts of the world as well as cooler high hills of sub tropical areas. 'An apple a day keeps the doctor away', this old adage focuses man's attention on the importance of apple in daily diet. The fresh apple fruit is a rich source of vitamins, minerals, proteins, fibre, carbohydrates etc. It is the most important fruit crop of temperate regions of the world and is cultivated on an area of about 5,641,856 ha with annual production of 60,964,800 metric tons. In India, apple is commercially cultivated in the states of Jammu and Kashmir, Himachal Pradesh, Sikkim, Arunachal Pradesh and Uttarakhand on an area of 240,000 ha with annual production of 17,93,000 metric tons. Its cultivation has revolutionized the socio-economic condition of hilly farmers in Himachal Pradesh and has become the number one commercial fruit crop and is being grown over 88,560 ha with annual production of 5,40,356 metric tons.

Apple is grown as perennial monocultures and hence, the pest and disease problems are entirely different and complex in nature. Such diseased situations have led to repeated and excessive use of chemical pesticides, which has resulted in the development of resistance in the pathogens and environmental pollution. Over 92 diseases (Table-1) have been reported to affect the apple crop resulting in reduction in yield, devaluation in grade, foliage loss and operational expenditure.

In the present context, wide spread epidemic nature of the summer diseases varying from region to region are becoming serious and economically important. It has been reported that the outbreak of these pathogens is due to changes in the winter weather conditions where under hot winters, late or less snow fall in the sub-temperate regions is consequently encouraging the summer diseases (leaf spotting fungi)

Table 1 : Apple diseases caused by biotic and abiotic factors.

S.No.	Disease / disorder	Casual organism
1	Alternaria leaf spot and storage rot	*Alternaria mali*
2.	American brown rot	*Monilinia fructicola*
3.	Anthracnose bull's eye rot and perennial canker	*Neofabraea malicortis*
4.	Bitter pit	Localized calcium deficiency
5.	Bitter rot	*Glomerella cingulata*
6.	Black pox	*Helminthosporium papulosum*
7.	Black root rot	*Xylaria mali*
8.	Black rot, frogeye leaf spot and canker	*Physalospora obtusa*
9.	Blister canker	*Nummularia discreta*
10.	Blister spot	*Pseudomonas syringae pv papulans*
11.	Blossom blast	Boron deficiency
12.	Blossom and shoot blast	*Pseudomonas syringae pv syringae*
13.	Blotch, leaf spot and twig canker	*Phyllosticta solitaria*
14.	Blue mold	*Penicillium expansum*
15.	Canker and fruit rot	*Botryosphaeria dothidea*
16.	Brown rot	*Monilinia laxa*
17.	Burr knot	Unidentified aerial root gall
18.	Calyx end rot	*Sclerotinia sclerotiorum*
19.	Cedar apple rust	*Gymnospororangium juniperi-virginianae*
20.	Clitocybe root rot	*Armillaria tabescens*
21.	Collar rot	*Phytophthora cactorum*
22.	Crown gall	*Agrobacterium tumefaciens*
23.	Cytospora canker and dieback	*Cytospora ambiens, C. leucostoma, C. personata*
24.	Dapple apple	Unidentified
25.	Eastern white root rot	*Scytinostroma galactinum*

26.	Fire blight	*Erwinia amylovora*
27.	Fish-eye rot	*Corticium centrifugum*
28.	Fly speck	*Schizothyrium pomi*
29.	Fruit cracking	Genetic
30.	Fruit russet	Frost, sprays, etc
31.	Gray mold	*Botrytis cinerea*
32.	Green crinkle	Unidentified
33.	Green mottle	Unidentified
34.	Hairy root	*Agrobacterium rhizogenes*
35.	Hawthorn rust	*Gymnospororangium globosum*
36.	Hollow apple	High temperature
37.	Hypochnus leaf blight	*Pellicularia koleroga*
38.	Internal bark necrosis	Low pH, nutrient imbalance
39.	Internal browning	Boron & calcium deficiency
40.	Jonathan spot	Unidentified
41.	Leptosphaeria canker	*Leptosphaeria Coniothyrium*
42.	Moldy core rot	*Alternaria alternata*
43.	Narrow leaf	Genetic
44.	Nectria twig blight	*Nectria galligena*
45.	Nematodes	
	a. Daggar	*Xiphinema americanum*
	b. Lance	*Hoplolaimus* sp.
	c. Pin	*Paratylenchus hamatus*
	d. Ring *Criconemella* spp.	
	e. Root knot	*Meloidogyne* spp.
	f. Lesion	*Pratylenchus penetrans*
	g. Sheath	*Hemicycliphora* spp.
	h. Spiral *Helicotylenchus* spp.	
	i. Stubby root	*Paratrichodorus minor*
	j. Stunt	*Paratricidorus* spp.
46.	Phoma fruit (Brook's) spot	*Mycosphaerella pomi*
47.	Phymatotrichum root rot	*Phymatotrichum omnivorum*
48.	Pink rot	*Cephalothecium roseum*
49.	Powdery mildew	*Podosphaera leuchotricha*
50.	Quince rust	*Gymnosporangium claviceps*
51.	Ring spot	Unidentified
52.	Rosette	Genetic
53.	Rough bark	*Phomopsis mali*
54.	Scab	*Venturia inaequalis*
55.	Septobasidium canker	*Septobasidium pedicillarum*

56.	Shoestring root rot	*Arimillaria mellea*
57.	Side rot	*Phialophora malorum*
58.	Silver leaf	*Stereum purpureum*
59.	Soft scald	Low temperature
60.	Sooty blotch	*Gloeodes pomigena*
61.	Storage scald	Injury to fruit surfaces by gases produced by fruit itself
62.	Sunburn	Sun injury to fruit
63.	Sunscald	Freezing of bark following high temperature in winter
64.	Superficial bark canker	*Pezicola corticola*
65.	Twig canker	*Monochaetia mali*
66.	Variegation	Chimaerial
67.	Blister bark	Graft transmissible pathogen (GTP)
68.	Chat fruit	MLO
69.	Chlorotic leaf spot	Apple chlorotic leaf spot virus
70.	False sting	Virus
71.	Flat apple	Cherry rasp leaf virus
72.	Flat limb	MLO
73.	Flute fruit	Apple stem pitting virus
74.	Fruit scurf	Virus
75.	Leaf pucker	MLO
76.	Mosaic	Prunus necrotic ringspot virus
77.	Platycarpa scaly bark	Virus
78.	Proliferation	MLO
79.	Pustule canker	Virus
80.	Rough skin	GTP
81.	Rubbery wood	MLO
82.	Russet ring	Virus
83.	Scar skin	GTP
84.	Spy lethal	Virus
85.	Star crack	GTP
86.	Stem grooving/graft union brown line	Apple stem grooving virus
87.	Stem pitting	Virginia stem pitting virus
88.	Tulare apple mosaic	Tulare apple mosaic virus
89.	Union necrosis & decline	Tomato ring spot virus
90.	Water core	Over maturity of fruit
91.	White heart rot	Various basidiomycetous fungi
92.	White root rot	*Dematophora necatrix*

incidence. In orchards the outbreak of summer diseases affect the crop directly in terms of productivity, quality and economical returns. The new emerging leaf-spotting diseases like apple scab, powdery mildew, *Marssonina* blotch, *Alternaria* spot, blotch and blight, *Phyllosticta* blotch and spot, *Phytophthora* leaf blight and fruit rot, *Coniothyrium* leaf spot, *Monilia* leaf spot and blight, *Mycosphaerella* leaf spot are perpetuating in the orchards. Total annual losses are about 28-36 per cent despite expenditure of approximately INRs 32000-40000/ha.

Total diseases affecting apple crop caused by fungi, bacteria, viruses and virus like organisms, nematodes, genetic and physiological disorders are 93 in the entire apple growing regions of India. All are identified and enumerated. Few major diseases causing heavy annual losses are described below.

Disease

A. Fungal disease

1. Apple scab

Causal Organism: *Venturia inaequalis*

Description: Scab may occur on leaves, fruit, petioles, and green twigs. Infections of the leaves and fruit are most common and obvious. Early season infections usually occur on the underside of the blossom cluster leaves because these are the first tissue surfaces to emerge from buds in the spring. Once the cluster leaves have unfolded and terminal leaves begin to develop, infections become evident on the upper surface of the leaves. Individual infections appear as rough, circular, and brown to dark olive-green spots (lesions), which often appear slightly fuzzy or velvety in texture. Lesions along the veins or margins often cause affected areas to distort or crinkle. Primary (ascospore) infections are usually limited to one or two distinct spots per leaf, whereas secondary (conidia) infections are often much more numerous. Secondary infections occasionally are so numerous that the entire surface of the leaf appears covered with scab, a condition commonly referred to as 'sheet' scab.

Lesions on young fruits initially resemble those on leaves but turn dark brown to black and become corky or scab-like with time.

Cells near lesions on young fruit may be killed, causing these regions to become deformed or cracked as they fail to grow and expand along with the remainder of the apple. Primary infections are usually limited to one or two distinct spots per fruit, often near the blossom end because it is upturned during the early stages of fruit development. Secondary infections are more frequent and numerous and the spots

may be clumped or grown together, particularly if the fruit is directly beneath a concentrated source of secondary spores such as an infected leaf. Secondary infections that occur in late summer or early fall are often numerous and relatively small in size, a symptom referred to as 'pinpoint' scab. Infections that occur just before harvest may be symptom-less at picking yet develop into 'storage' scab lesions after harvest.

Disease Cycle

The fungus *Venturia inaequalis* causes apple scab, which is now endemic to all apple- growing countries of the world and is particularly severe in areas having high relative humidity and rainfall.

In winter, the scab fungus over-winters in infected leaves fallen to the ground. During autumn, the fungus begins to form tiny fruiting bodies, which are embedded in the leaves near the surface. Sacs filled with the primary or spring spores of the fungus start to develop within the pseudothecia by late winter or early spring. The ascospores continue to develop and mature as spring progresses. A few spores are usually mature at the time of bud break (green tip), and maturity progresses slowly until about the tight cluster stage of blossom development. After this time, the percentage of mature spores begins to increase rapidly whenever temperatures are favourable for tree growth. Most ascospores have matured by the end of bloom. Mature ascospores are discharged into the air during periods of rain. In daylight, discharge usually begins within 30 minutes after the start of the rain and is largely completed within 3 to 6 hours. When rainfall begins at night, discharge is often delayed until daybreak, although significant night discharge can occur under some conditions. The number of spores discharged during any one rain is determined by both the size of the potential ascospore "crop" for the season (how many leaves were infected the previous year) and the percentage of these spores that have matured since the last discharge. Ascospore discharge usually peaks in the time from pink through bloom, and nearly all ascospores are discharged within 1 or 2 weeks after petal fall.

Ascospores are blown to nearby trees by wind currents, and they germinate in a film of water on the surface of leaves and fruit. If surface wetness continues long enough at prevailing temperatures, growth from the germinated spore penetrates and infects the organs just beneath the outer cuticle. Typical lesions, each bearing tens of thousands of secondary or summer spores, appear about 9 to 17 days later depending on temperature, although long periods of low humidity can delay their development. Conidia are dispersed by splashing rain throughout the rest of the season and are capable of causing new infections. Due to numerous additional conidia produced on each new lesion; repeated secondary infections have snowball or epidemic effect on disease development.

Incidence of infection is affected by the age of leaves and fruit; young leaves generally are most susceptible dunks in 1 to 5 days after unfolding and become completely resistant from the time they finish expanding until shortly before leaf drop in the autumn. Fruits are highly susceptible until about 3 to 4 weeks after petal fall, but much longer wetting periods are required for infection to occur after this time. Precise requirements for infection of mature fruit are not known, but limited data indicate that wetting periods must last at least 48 hours for significant infection to occur immediately before harvest.

2. Powdery mildew

Causal Organism: *Podosphaera leuchotricha*

Description: The disease infects leaves, blossoms, green shoot, and the fruit. On the leaves, the fungus can appear as felt-like patches or as a solid mat on the surface, particularly on the undersurface of the leaf. Initial infections on the underside of the leaf may cause chlorotic patches or spots to occur on the upper side of the leaf. This symptom, however, is not unique to powdery mildew so inspection of the underside of the leaf is necessary to confirm that powdery mildew is the cause of these symptoms. Under favorable environmental conditions, the disease spreads over the entirety of the leaf and progresses down the petiole on to young, green shoots. Infected leaves tend to crinkle, curl, or roll upwards along the edges giving them a narrow appearance. If the infection is severe, infected leaves will usually drop prematurely during the summer.

Infected terminal shoots are stunted and the leaves along the shoots appear as described above. Shoot infections are the typical result of overwintering infections of the fungus in buds. When the terminals push in early spring, the fungus grows along with the new succulent growth and infects this tissue immediately. These shoots may be killed outright in the spring, or may survive throughout the season and die in late fall or winter. The initial growth of the fungus on newly infected twigs appears powdery white but eventually turns a darker brown. Small black fruiting bodies, called cleistothecia, can appear in the mycelial mat and function as the source for sexual reproduction and the eventual production of ascospores.

The blossoms, petals, sepals, receptacles, and peduncles may become infected and covered with the fungus. Blossom infections are less common but are important because infected blossoms will either fail to set fruit or produce small, stunted and/or russetted fruits. These fruit are unmarketable for fresh market consumption.

Disease cycle

The powdery mildew fungus overwinters as mycelium in infected buds or as cleistothecia on the surface of infected twigs. Infected terminal buds are more susceptible to winter injury than healthy buds, typically bud break in the spring is 5-8 days later than healthy buds, and are more susceptible to spring frost than healthy buds. In fact, healthy buds may survive at temperatures 2-10°C colder than infected buds. As a result, many infected buds will not survive through a cold winter. This is important to keep in mind because a hard winter can dramatically reduce both disease pressure and the need for control measures during the subsequent season. The survival rate of infected buds is less than 5 per cent when temperatures drop below –24°C and, although not well studied, it appears that temperatures around –12°C will kill the mycelium in infected buds and allow the bud to produce healthy leaves. The cleistothecia, which form the sexual stage of the fungus and produce ascospores, apparently do not play an important role.

Mycelia in infected buds produce conidia to initiate primary infections and these infections can be found as early as the tight cluster stage. There is often an abundance of susceptible tissue for the conidia to infect because infected buds break later than healthy buds. Conidia are disseminated by wind and can infect young leaf tissue, blossoms, and fruit. Leaves become increasingly resistant to infection as they age and become nearly immune once they have matured, although infection can occur through injuries on older leaves. Symptoms may develop as early as 5 days after infection. Numerous secondary cycles can occur under favorable conditions and, like many powdery mildews, cooler temperatures rather than relative humidity drive early secondary infections. Like other powdery mildews, *P. leucotricha* exhibits a diurnal periodicity in that the highest concentration of airborne conidia is found from midday to early afternoon. Infections that result in fruit russet occur primarily during the pink stage of bud development.

Apparently, infection of lateral and fruit buds occurs within 1 month after they are formed. The infections remain latent until bud break the following spring where they will serve as the initial source of inoculum. The lateral buds are susceptible to infection longer than the terminal buds, however, it is the terminal buds that are the likely source of overwintering of the fungus as infection can be greater than 50 percent by terminal bud set.

3. Apple blotch

Causal Organism: *Marssonina coronaria* and *Alternaria mali*

Description: Since 1995, apple plants manifested another equally important disease causing pre mature defoliation in all apple-growing states viz. Himachal Pradesh, Jammu and Kashmir, Uttarakhand and North- east in recent years. The disease is caused by fungus, *Marssonina coronaria* and *Alternaria mali* resulting in mid-season defoliation. Its manifestation has been observed on all the commercial cultivars grown in India. Disease appears both on leaves and fruit. Dark brown patches develop on upper surface of leaves varying from 5-10 mm diameter which are typical blotch symptoms. Various small and large blotches depending on the climatic conditions after the initiation of disease coalesce to form large dark brown blotches and the surrounding areas turn yellow. The yellowing advances towards petiole end and induce mid season defoliation. Economic losses from the disease have yet not been documented as the malady is of recent origin, but it reduces the fruit size and delays ripening. The fruits are poorly coloured and lack flavour. Irregular necrotic blotches on mature leaves characterize the disease. The blotches develop suddenly and increase rapidly during June-July and cause yellowing and defoliation. The mid-season defoliation may be responsible for poor growth and development of fruiting buds and result into low productivity in ensuing year. Infection of the fruit appears in the harvested fruits and dark brown depressed spots are most common on 'Golden Delicious'.

Disease cycle

The fungus overwinters in the form of apothecia on fallen leaves. Primary infection is caused by ascospores produced on the overwintered leaves. The ascospores discharge begins during spring season but causes infection at a particular leaf age stage. The disease first attacks plants growing in shady and high moisture conditions. Air temperature, light intensity and soil moisture are the predisposing factors to trigger the disease outbreak, however, severity index increases with the maturity of the leaves. The disease is severe when a cool, rainy period of 4 to 5 days proceeds several hot sunny days in the orchard. The inoculum of the disease get released at the pink bud stage but it causes the infection only at a particular leaf maturity stage when plant physiological processes start producing ethylene, the ripening / ageing hormone. The pathogen therewith the hormonal changes induce yellowing and abscission of the leaves and defoliation occurs. The disease process key on the rate of infection conditions is developed for this new disease on apple plant to monitor disease. It therefore concludes that the blotch started its symptoms on the upper side of the leaf first in the 2nd week of June irrespective of the weather conditions. However, the disease symptoms on leaves followed by temperature above 20^{0}C and relative humidity above 70 per cent continuously for 5-6 days, induce yellowing and defoliation process. Prolonged hot and dry weather affect the primary infection and disease conditions adversely. Under favourable weather conditions the leaves start defoliation within 3-5 days whereas under interrupted/unfavourable conditions the infected leaves could retain up to 3 months.

Disease monitoring system for the farmers

A seven years infection conditions were monitored and laid out to a host-physio-epidemiological model wherein disease outbreak and subsequent symptoms to the defoliation were described. To make the disease monitoring system more purposeful and applicable, the apple growers are being advised to follow the infection pattern based on their orchard sites. The first monitoring step is to average fully open leaves as zero days and adds 65 to 68 days including 6-28 days incubation days; the date would be the first infection day and date of first preventive spray as well. It would generally be first week of June in most of the orchards. However, similar infection criteria in orchard at an elevation of 8146ft were met in first week of July. The farmers are therefore provided with the information to develop an annual self-record of their orchards and match the sequences with the standard method. The information chart is already provided to way out to monitor and manage the disease effectively in minimum sprays.

4. Canker disease complex

Causal Organism: Species of *Sphaeropsis, Corticium, Botryosphaeria, Coniothecium, Nummularia and Chondrostereum*

Description: With the increase in area under apple cultivation along with growing of certain susceptible varieties, the numbers of new canker diseases have appeared. Among them stem and branch cankers cause huge losses through girdling of branches, limbs and dieback of twigs resulting in death of plants. The bark beneath the rough exterior becomes dry, hard and tough. Cankers develop lengthwise more rapidly and become elliptical on limbs and branches. Cankers may extend to more than a meter in length and may completely girdle the affected parts. Perennial cankers produce numerous pimple-like protuberances and fruiting bodies of secondary fungi over the rotted or decayed wood. Various cankers induce varied type of symptoms depending on the fungus and climatic conditions of the orchards. It is very difficult to estimate the loss incurred due to canker since many factors are involved. The canker can be defined as diseased

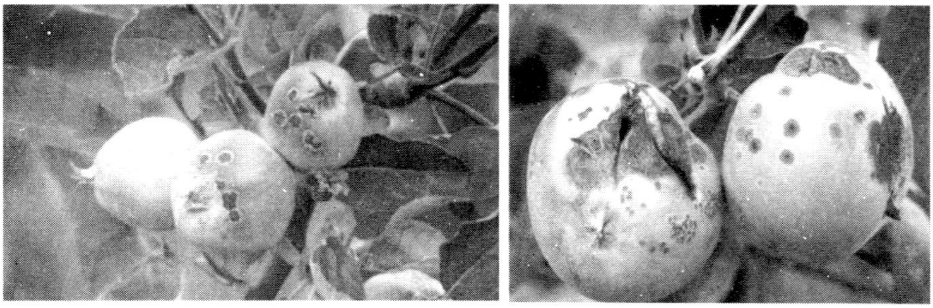

Apple scab

Powdery mildew

Apple blotch

Canker disease complex

area on the stem or branch usually well defined which often results in death of bark within the infected area. Cankered area may be large or small and may extend to wood or remain more or less superficial. Repeated callusing is necessary before a lesion can be classed as 'canker'. Faulty pruning over the years, adverse weather conditions, and wound/injuries caused while undertaking agronomical practices and also by insect-pests, favour canker development. Other factors affecting the prevalence of canker are type of cultivars, tree age, site and altitude of orchard. Severity of cankers is more a matter of concern in marginally situated apple orchards (900-1,200 m.a.s.l.), where the chilling hours (>1200 hr when temperature must be below 7°C) requirement is not met with completely. Cankers occur in all apple-growing areas of the world including India. In Himachal Pradesh, thirteen different cankers have been recorded so far and six of them namely; smoky blight (*Sphaeropsis malorum*), pink canker (*Corticium salmonicolor*), stem canker (*Botryosphaeria dothidea*), stem black (*Coniothecium chomatosporium*), nail head canker (*Nummularia discreta*) and silver leaf canker (*Chondrostereum purpureum*) were found to be most prevalent.

5. White Root rot

Causal Organism: *Dematophora necatrix*

Description: White root rot in apple trees has been increasingly imposing problem and resulting in the death of plants. This soil borne fungus has a very wide host range of around 173 plant species. Further, death of a tree, due to root rot in an orchard does not mean the end of losses, instead the incitant being soils overlying stony sub-soils. Of the several factors examined in Himachal Pradesh for their possible role in the longevity and survival of the root rot pathogen, clay loam soil was reported better than clay soil for its longer survival at all temperature ranges and soil moisture regimes. The silty loam soil did not favour the survival; *D. necatrix* survived for 36 and 737 days more than in clay and silty loam soils under excessive soil moisture at 17°C respectively. Temperature has been known to play an important role in the survival and spread of the pathogen. *D. necatrix* has been observed tolerating wide temperature range (2-23°C) but 17°C was an optimum temperature level. Of the two factors i.e. temperature and soil moisture, the latter plays more decisive role in increasing the longevity of the pathogen than the former. In excessive soil moisture conditions and a wide range of temperature the pathogen survived for 800-963 days more than at normal and dry soil conditions.

The prevalence of white root rot in Himachal Pradesh is limited to high hill soil texture which varies from loam to clay loam. The soil reaction is slightly acidic and this zone is ideally suited for apple cultivation. It receives an average of 100-140 cm rainfall per annum. The rainy season in the apple zone is during June to September, when also a higher incidence and severe symptoms of root rot are observed, indicating that the disease spread and severity are related to excessive soil moisture regimes. The higher elevation range was observed favourable for the spread of *D. necatrix* being high rainfall area and low temperature conditions prevailing for long durations. Besides soil, pH also seems to have a strong relationship with the survival of the pathogen flourishing best in 6- 6.5 pH range. Incidentally the larger area under apple cultivation in Himachal Pradesh also falls within this pH range.

However, it seems that the incidence and survival of the pathogen of apple root rot is dependent on physio-environmental conditions, viz., soil type, soil reaction, soil temperature and soil moisture, which are available altogether in the apple areas recorded for high incidence of root rot disease in Himachal Pradesh.

6. Collar Rot

Causal Organism: *Phytophthora cactorum*

Description: The disease infection starts from the collar region and spread mostly to the underground parts and the above ground stem portion is also infected in highly susceptible scion cultivars. Bark at the soil level becomes slimy and rots resulting in cankered areas. The wounds are irregular in outline but usually roughly oval which extend rapidly, often resulting in girdling of the tree. The attacked trees are recognized by chlorotic foliage with red colouration of veins and margins.

7. Leaf spotting fungi

Causal Organism: Species of *Alternaria, Marssonina, Coniothyrium, Mycosphaerella, Phyllosticta, Monilia and Phoma*

Description: Leaf spots are important aerial fungal diseases affecting all apple growing regions invariably for last few years and causing heavy losses in the productivity of the plant due to premature defoliation and reduction in photosynthetic activities. Thus leading to reduced vigour and viability of the plant and in turn lowering the fruit production every year. These foliar pathogens also harm the fruit by causing spots leading to rotting of fruits. Inspite of having direct losses on fruit quality and productivity, the impact of these pathogens is assumed to be insignificant due to lack of much scientific work on their biological and epidemiological parameters as well as management schedule. Till date there is very little information available on the individual losses caused by various leaf spot pathogens but the cumulative losses are very high and cause an overall decrease in fruit productivity. Leaf spots are one of the major factors responsible for declining productivity and increased recurring expenditure of apples. Few major leaf spotting diseases are making their presence viz., *Alternaria* (spot/blotch/blight) *mali, Marssonina* (blotch) *Coniothyrium* (leaf spot) *pyrinum, Mycosphaerella* (leaf spot) *pyri, Phyllosticta* (blotch) *pirina, P. mali* and *P. pyrina, Monilia* (leaf blight) *mali, Phoma* (leaf spot) *pomorum, P. macrostoma, P. glomerata* and *P. exigua* and causing heavy losses.

The blemishes on the leaves resulting in chlorotic or necrotic lesion are generally termed as a leaf spot. These spots start appearing with the emergence of leaves however severity and damage depends on climatic conditions and topographical locations. In recent years the cumulative loss to the productivity has been estimated up to 28 per cent.

8. Fly Speck

Causal Organism: *Zygophiala jamaicensis* and *Mycrotheria* spp.

Description: Symptoms of flyspeck on apple are approximately described by the name. Black spots are formed on fruits. Colonies are characterized by well-defined groups of 6, 8 or sometimes 50 shiny, black, superficial pseudothecia on the surface of the fruit. This disease is found abundantly during rainy days and in less windy regions. Sooty blotch and flyspeck colonies commonly occur on the same fruit, but the colonies are mutually exclusive.

9. Sooty Blotch

Causal Organism: *Gloeodes pomigena*

Description: Sooty blotch colonies appear as shades of olive green on mature fruit. They vary from discrete, almost circular colonies to large colonies with diffuse margins which are sooty in appearance. Large amorphous colonies that cover large portions of the fruit surface result from secondary spread on the fruit. The fungus is found on the upper surface of fruit, which can be easily removed by rubbing leaving behind slight depression. No permanent spot is found on the fruit surface. This disease is found on late maturing varieties and under storage conditions. The fungus survives from one season to the next as mycelium and pycnidia on infected twigs of apple and numerous forest trees. The fungus is dispersed as chlamydospores or conidia during rains in the spring and early summer. Most conidia are liberated by early summer, but extensive secondary spread by chlamydospores and mycelial fragments occur throughout the seasons. The infection can also be seen on new twigs, on which it remains active. During the months of May and June, low temperature and rains are necessary for disease development. For the fungus development 18.3°C is favourable. The disease is more prevalent in low and shady area of the orchard.

10. Storage rot

Causal Organism: Species of *Penicillium, Botrytis, Monilinia, Trichothecium* and *Alternaria*

Description: Apples are often stored for 6-9 months after harvest and are subject to attack by numerous post harvest pathogens. A 1931 list of fungi causing apple storage decays included 90 species have been described since that time. Despite modern storage facilities, losses of 15-20 per cent have still been reported. Major losses were attributable *to Penicillium, Botrytis, Monilinia, Trichothecium* and *Alternaria*. In addition, various causal organisms are encountered when apples are marketed under hot conditions in India. Commonly occurring storage rots are:

a. Blue mold

Casual organism: *Penicillium expansum*

Description: Blue mold, also known as soft rot wet rot, is the most important post harvest disease. A good diagnostic for blue mold is the characteristic earthy, musty odour associated with the decay. The decayed flesh is soft and watery and separates easily from the healthy tissues. The common name for this disease derives from the tufts of bluish or blue-green spores that sometimes appear on the surface of the decay or along the edges of wounds. Blue mold infection can originate from the wounds, stem-end invasions, or core rots. *Penicillium* spp. known to cause blue mold can be isolated from most of the orchards soils, but disease is rare in the field except on fruits fallen on the ground. Most fruit infections occur when airborne or waterborne conidia enter wounds incurred during harvesting and handling.

b. Gray mold

Causal organism: *Botrytis cinerea*

Description: Gray mold is important storage rot on apples particularly at cold-storage temperature. Gray mold lesions first appear as pale tan areas without sharp margins. Infections may originate from wounds, stem punctures, or the stem or calyx end of the fruits. As the decay enlarges, the older portion of the decay may turn darker brown, but the edges remain a pale colour. On red apples, dark areas may persist around lenticels even after the fruits are completely decayed. On less matured fruits, the rotted tissue is firm and does not separate easily from healthy tissues. Completely rotted fruit often appear baked and decayed flesh has a sweet, cider-like odour. Under high humidity, white or gray-white mycelium may develop on the surface of decayed areas. Sporulating mycelium appears gray, but little sporulation occurs at cold-storage temperature.

Gray mold rot is seldom seen in the field, although *B. cinerea* is a common saprophyte on decaying organic matter on the orchard floor. The fungus causes dry eye rot in the field, but that disease is unimportant in the epidemiology of post harvest gray mold. Wound or injuries are the primary infection courts for the initiation of gray

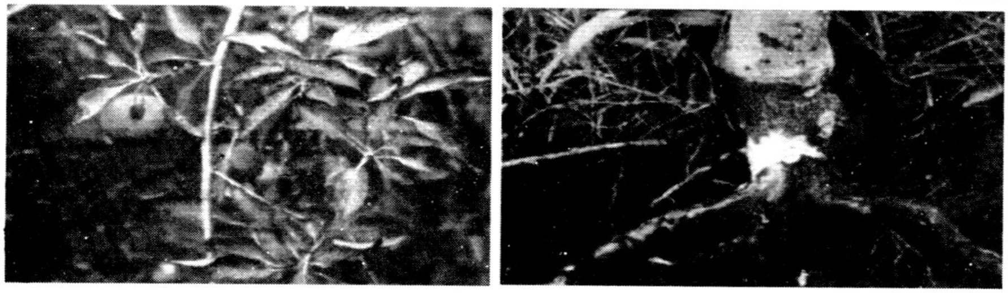

White root rot

Collar rot

Fly speck

Sooty blotch

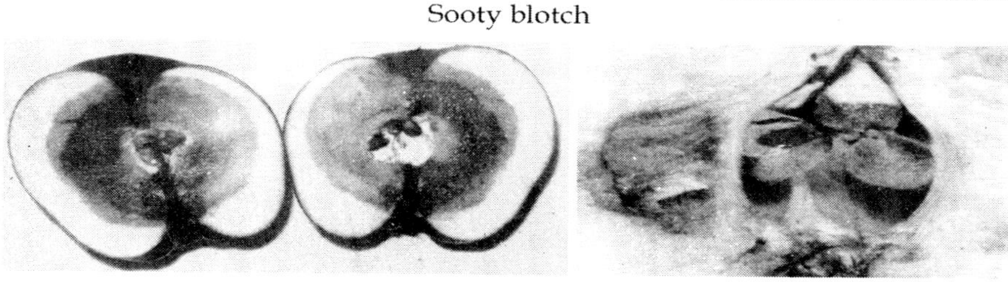

Moldy core rot

mold. Conidia of *B. cinerea* are carried in the orchard soil or produced on the other organic debris. Fruit that decay during storage produce additional inoculum. Inoculum once established in a wound, *B. cinerea* can quickly spread from infected fruit to adjacent healthy fruit during storage.

c. Moldy core rot

Casual organism: Species of *Alternaria* and *Trichothecium*

The occurrence of apple fruit drop and rotting of seed core/cavity/ fruits is called Core rot of apple. However, in the present instance the disease is Moldy core rot or dry core rot having fungal growth around the seed cavity. Other such core rottings are called Wet core rot caused by different fungi. The Moldy core rot is also caused by many different species of fungi, among them *Alternaria* spp., *Stemphylium* spp., *Cladosporium* spp., *Coniothyrium* spp. and *Trichothecium* sp. are most common and are always present as facultative parasite or hyperparasite on the apple plant surfaces. However, the incidence of the present Moldy core rot is due to *Trichothecium roseum and Alternaria spp.* (based on microscopic examination and characteristic pinkish growth of the fungi). Wet core rot appears after harvest as a soft, internal rot in storage.

This disease is characterized by infection within the locules (the cavity of the ovary or seed cavity), without penetration into the fruit flesh. External symptoms on the fruit are quite subtle, and typically the disease goes unnoticed until the fruit are cut open. External symptoms may include premature ripening due to infection on the seed and infected fruit may drop from the tree.

The fungi responsible for this disease colonize the flower parts as soon as the blossom opens. The fungi then enter the developing fruit through sinus opening in the calyx. Such fruits after setting remain open with calyx side in upward direction and petals are attached to the calyx end. This phenomenon generally invites fruit to infection. The fungal mycelium or spore enter the seed region, which intially remains latent due to no seed formation but after seed development takes to brown colouration, the growth of the fungus colonizes the seeds. The rotting starts with the induction of sugar/

sweetness in the fruit. Weak or shrivelled seeds are more vulnerable to core infection. The intensity of rotting and subsequent dropping are affected by the sugar in the fruit thereby effecting early maturity first in early maturing varieties and later on all the Delicious cultivars. The infected fruits ripen earlier, drop and are more sweet to eat.

Moldy core is primarily a problem during years having wet flowering period or in years when heavy rains follow dry weather in early summer in late summer. Wet weather during bloom may cause conditions favourable for the fungi to produce spores. The associated fungi are always present on the plant surfaces, which drift over to the floral parts by water or air but the chances of such infection are less. The bees/insects in the process of pollination can act as carrier of spores to the floral parts where infection takes place to the internal parts of the flowers. Apple cultivars vary in their susceptibility to moldy core. Early maturing cultivar like Red June and Tydeman's-Early-Worcester had moldy core rot and dropping abnormally.

B. Bacterial Disease

1. Crown Gall

Causal Organism: *Agrobacterium tumefaciens*

Description: Galls varying in diameter from several millimeters to 5 to15 cm on older trees are located on the crown and roots, often at pruning wounds, and occasionally on stem. The young gall are pale, soft, smooth, composed of light coloured frosty tissue resembling fresh callus growth, but gradually becomes dark, hard and deeply fissured. Affected trees may become stunted, produce poor root systems, small and chlorotic leaves, and blackening of stem tissues near galls.

2. Hairy Root

Causal Organism: *Agrobacterium rhizogenes*

Description: Hairy root symptoms on young apple tree are characterized by an extensive proliferation of adventitious roots, singly or in clusters. After 2 to 3 years, affected trees become weak and fruit yield is reduced.

C. Viral/Viroid Disease

1. Apple Mosaic

Causal Organism: Apple mosaic virus (ApMV)

Description: In the spring, pale yellow to cream-coloured areas develop on expanding leaves of infected apple trees. These areas may appear as small spots, flecks, irregular blotches, vein-net patterns, line patterns, or bands along major veins. As the season advances, they turn from pale yellow or cream to a lighter chrome yellow or white and then soon become necrotic. Severely affected leaves abscise prematurely.

2. Apple Chlorotic Leaf spot

Description: Apple chlorotic leaf spot virus (ACLSV) the type species of *trichovirus* genus is probably the most common virus of apple and pears. It occurs naturally in apple, pear, quince, plum, apricot, peach and cherry. Chlorotic leaf spot virus induces a wide range of symptoms in various *Malus* species.

The symptoms appear generally on leaves, fruits and trunk and their severity depends largely on plant species and virus strain. The symptoms include translucent or chlorotic leaf spots with asymmetric leaf distortion, irregular diffused chlorotic rings and line pattern on leaves of reduced size, stunting, terminal die back of some clones, pitting of the xylem and inner bark necrosis. Several cultivars can be used as virus indicators owing to their unusual sensitivity to chlorotic leaf spot virus.

The virus may be classed as one of the latent apple viruses and does not produce symptoms in most apple varieties. It gives rise to stunting, leaf distortion and chlorotic spotting in most of the apple varieties and reduction in tree vigour in Golden Delicious.

Disorder

Physiological Disorders

a. Bitter Pit

Description: Pits are often associated with the terminals of branched vascular bundles. The lesions are small, brown, dry, slightly bitter-tasting, and about 3 to 5 mm in diameter. Internal lesions can

Crown gall

Apple mosaic

Apple chlorotic leaf spot

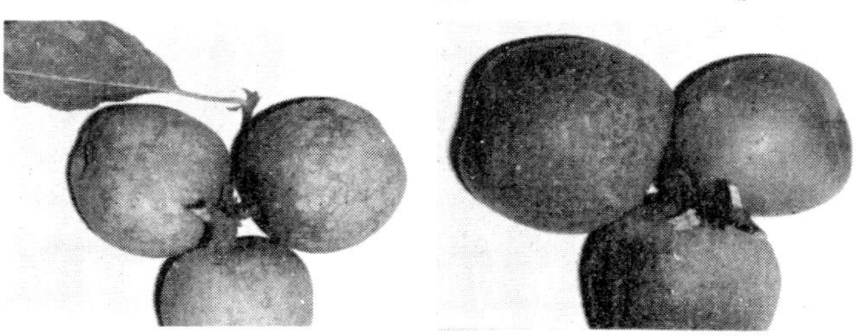

Russeting

Hairy rot

occur any where in the tissue from the core line to the skin, but they are more common just below the skin. In badly infected fruit, they may coalesce into blotches. When first visible, the skin of the fruit over the affected area appears water-soaked. Later the spots become purple on blushed skin, dark green on less-coloured areas. The skin over the pits gradually becomes brown or sometimes black. The main cause is thought to be a mineral imbalance in the apple flesh.

3. Russeting

Description: Russet is usually initiated early in the life of the fruit, about 30 days after anthesis, which coincides with the greatest increment of tangential growth, which occurs 23 days after anthesis based on equatorial cross-section measurements.

Russeting occurs in various forms, ranging from the rough scabrous type, which covers on full fruit, to light netted. Severe damage produces cracks on the fruit and ultimately fruit splits. Under storage conditions such fruit shrivels. A number of structural, physiological and environmental factors affect russet development and ultimately fruit splits. Under storage conditions such fruit shrivels. A number of structural, physiological and environmental factors affect russet development.

Causal Agents

Internal factors: The amount of russet is associated with the position on the fruit surface, and the position of the fruit in the cluster. Fruit cracking and russeting are both associated with a disintegration or breakdown of the cuticular layer. Cultivars with thin cuticles, or areas of fruit having thin cuticle are more subject to russet. The vigorous growth of tree is conducive to reduce incidence of tree is conducive to reduce incidence of cracking. The incidence of fruit russet is also related to impaired foliage and greater exposure to weather conditions - especially at an early stage of fruitlet development. Early mildew infections on blossoms and fruitlets are direct cause of the fine web-like russeting.

Prospects of Bio-pesticides and Microbial Organisms in Apples

Bio-pesticides and microbial organisms are the new development in insect pest and diseases management as an important IPM tool. The practice is the only resource to conserve bio-diversity comprising of beneficial pollinators / parasites / predatory insects, to obtain higher yields of crops through superior control of insect-pests and diseases of crops and to avoid losses caused by the presence of toxic residues of chemicals in crop management. Therefore, bio-pesticides and biocontrol agents in the country are in increase as these are advantageous due to their eco-friendly, target specificity, suppress resistance, reduced number of applications, yields and quality improvement, higher acceptability and value of produce for exports and economic returns. Botanical pesticides / bio-pesticides: are plants derived products known and utilized by man since ancient times. The chemicals synthesized by them repel approaching insects, deter feeding and ovi-position on the plants, disrupt behaviour and physiology of insects / pathogens in various ways and even prove toxic to different developmental stages of many insects / pathogens. Various plant species are known to carry medicinal or anti-microorganism actions. The most known and exploited plant for extracting insecticidal properties from every parts of the plant is Neem (*Azadirachata indica*).

Neem-based pesticides

Neem-based pesticides are biodegradable, emulsifiable, effective against a broad spectrum of insect pests, eco-friendly, harmless to plants, non-phytotoxic, maintain the stability of soil fertility, helps in growth regulation (eradication of pest), results in greater production, compatible in nature, easily applicable, economical, have better shelf-life compared to other bio-pesticides and are based on micro emulsion technology. Commercialization of neem-based pesticides (Neemarin, Achook, Aum Neem Plus, etc.) is expanding at a rapid rate because a number of insects, plant pathogens, nematodes and mites are being controlled by them. Nearly one hundred products are either being marketed or are awaiting commercialization in India. Due to neems legendry insectrepellent and medicinal properties, it has been identified as the most promising of all plants by the National Research Council, Washington, USA. All parts of the neem tree possess insecticidal activity but seed kernel is the most active. Neem based pesticides are used to control different pests (hoppers, thrips, stem borers, fruit borers, fruit flies, leaf minors, flea beetles, fruit sucking moths, leaf webbers, tea mosquitoes, etc.) in different fruit crops (grapes, pomegranate, banana, citrus, oranges, apple, mango, ber, etc.). Neem extracts with various solvents especially ethanol have been found to exhibit activity against insect-pests. Concentrations ranging from 0.001 to 0.4 per cent of various neem seed kernel (NSK) extracts have generally been found to deter feeding of most of the insects evaluated so far. The growth inhibitory affects of neem derivatives results in various developmental defects and even mortality in insects. Neem products also affect insect vigour, longevity and fecundity. Azadirachtin based 1500 ppm, 3000 ppm products are used for control of aphids, white flies, beetles, caterpillars, shoot and fruit borers, etc. In apple Neemarin and Achook commercial products were tested and validated against blossom thrips, which were found effective, equal to chlorpyriphos insecticide but safe to the non-target insects.

Advantages of Neem-based pesticides over chemical pesticides

- Compatible with the plant system- Neem-based pesticides being a product from nature when sprayed on the plant does not create any adverse effect on the same, which would otherwise result in the loss of crop.
- Soil fertility - Due to abundant usage of chemical fertilizers and pesticides, the soil loses its fertility thus resulting in yield loss. Neem-based pesticides on the other hand restore the fertility of soil.
- Eco friendly - Neem-based pesticides are degradable in 1 month and do not interfere with the physiology of plant. Unlike chemical pesticides and neither does it remained on the plant surface nor in the crop harvested. So there is no pollution caused in the atmosphere. Also it does not toxicate the human body.

- No residues - Owing to the degradability of Neem-based pesticides, the presence of the same in the plant as well as in the soil is marked off. Thus assuring chemical free food.
- No pesticide resistance - More and more usage of chemical pesticides helps the pests to develop resistance. Thus in due course of time, chemical pesticides cease to have any effect on pests infesting crops. On the other hand pests are unable to develop resistance to Neem-based pesticides

Microbial pesticides: Bio-agents: Biological control is the reduction of the amount of inoculum or disease producing activity of a pathogen accomplished by one or more organisms other than man. Basically, the role of a biological entity in suppression or destruction of pathogen / parasite constitutes biological control. It is based on the natural phenomenon that every biological entity has its adversary in nature which keeps its population under check. There are three approaches to biological control of plant pathogens: biological control of inoculum, biological protection of plant surfaces, and cross protection / induced resistance. Due to cosmopolitan nature of microorganisms and easy cultivation, most of the genuses have been utilized as agents for biological control of a number of important plant pathogens. Microbial pesticides are prepared using micro-organisms such as fungi, bacteria, protozoa and nematodes. Among various reported microbial agents tried and exploited for bio-control properties against cultivated crop diseases and insect-pests, the Trichoderma (fungus) and*Bacillus*(Bacteria) species are most promising and broad-spectrum to wide range of pathogenic species.

Among the known genus, *Trichoderma* (*T harzianum* and*T viride*) is one of the most commonly used bio-control agents. It is known to have tremendous potential as biocontrol agents when the selection of the species is done on the basis of antagonistic traits as well as other behavioural traits. The high degree of ecological adaptability shown by strains within the genus *Trichoderma* is reflected in the fact that these fungi are common in soils all over the world, under differing environmental conditions, and living on various substrates. This considerable variation, coupled with their amenability to cultivation on inexpensive substrates, makes *Trichoderma* isolates attractive candidate for a variety of biological control applications.

Trichoderma

Trichoderma is a very potential fungus for the control of many plant diseases, specially soil and seed borne diseases as it either directly parasitizes the pathogen or competes with other micro-organisms and pathogens for food requirements. Phytotoxic compounds produced by *Trichoderma* spp. are directly antimicrobial in nature and show antifungal and herbicidal properties. It produces a number of enzymes such as cellulases and hemicellulases, chitinases and glucanases. The enzymes chitinases and glucanases in their multiple forms are responsible for fungitoxic activity. Trichoderma spp. controls *Botrytis*, *Sclerotinia* and *Phytophthora cactorum* on apple. Even *Armillaria* sp. (a root pathogen) entry in the wounds can be prevented by *Trichoderma* application. *T.viride* application controls silver leaf canker of apple. Similarly, *T. koningii* also shows inhibitory effects against European canker. *T. harzianum*, *T. viride* along with soil solarization have been found to protect the apple plants from white root rot pathogen. Repeated applications of these antagonists enhance their efficacy against the pathogen and effect is long lasting.

Advantages of *Trichoderma* applications

- Provides improved disease control
- Helps in resistance build up
- Results in plant growth promotion (enhanced germination, rapid rate of flowering, increase in height weight, increased growth)
- Exhibit rhizosphere competence
- Saves the environment from pollution by hazardous chemicals
- Results in no residual effects

The commercial preparations of these fungi are available in the market in different commercial forms (Bioderma), which are easy to use. The low doses of fungicide combined with antagonistic application can prove to be more effective as the pathogen initially gets weakened and then subjected subsequent attack by the antagonists.

Bacillus

Microbial pesticide based on bacteria has gained much importance because various formulations are prepared commercially and the products (Biolep, Bioasp, Delfin, Dipel, Halt, Spicturin, Hil Btk, Floorback, 'Bt', etc.) are available in the market. These pesticides are effective, environment friendly, and safe to human beings, beneficial parasites I predators and other vertebrates. The 'Bt' based products are mainly giving better control of lepidopterous pests and are marked under various trade names such as Biobit®, Dipel®, Javelin®, etc. Trade names for these products (based on 'Bt'var israelensis) include Skeetal®, VectobaC®, etc. Raven® provides enhanced control of beetles and caterpillars that attack members of Solanaceae family. Additives (sticking or wetting agents) often are useful in a 'Bt' application to improve performance, allowing it to cover and resist washing.

Bacteria particularly *Bacillus* spp., have received maximum attention as microbial control agents. *Bacillus* produces various toxins that results in control of pests (*Helicoverpa*, *Plutella*, caterpillars, looper and semi-loopers, bollworms, Spodoptera, fruit and shoot borers) of agricultural crops and fruit trees by generally invading and mulfiplyinq in an insect and spread to infect other insects. *Bacillus subtilis* shows control over European canker and collar rot of apple.

Further Reading

Takur V. S. and Xu X (2004) Integrated Apple Orchard Management: A technology Book for Farmers. Published under Europian union International Coorporation Project and National Agricultural Project. 210.

□□□

Chapter 4

DEVELOPMENT OF IMMUNOLOGICAL KITS FOR EARLY DETECTION OF FOLIAR FUNGAL PATHOGENS OF *CAMELLIA SINENSIS*

S. RANA, R. DAS BISWAS, R. SOM AND B. N. CHAKRABORTY

Immuno-Phytopathology Laboratory, Department of Botany, University of North Bengal, Siliguri-734013, West Bengal

Introduction

Tea [*Camellia sinensis* (L.) O. Kuntze] leaves, besides providing the most healthy drink, are now-a-days being exploited for other innovative aspects like extraction of active principles that give protection from cancer and other diseases (Chakraborty and Chakraborty, 1998). Therefore, the importance of this crop for human beings cannot be overemphasized. Perennial habit of the tea plant, peculiar cultural conditions and warm humid climate of the tea growing areas are highly conducive for disease development (Hajra, 2001). A large number of pathogenic organisms are available in this ecological niche. Leaf diseases are more important due to the obvious reason that tea plant is cultivated for its young succulent leaves (Muraleedharan and Chen 1997). Early detection of these foliar fungal pathogens is often critical for success in eradication of the disease. The use of serological assays for detection and diagnosis of plant diseases is rapidly increasing (Chakrabory and Chakraborty, 2003; Gawande *et al*, 2006; Kennedy *et al*, 2000). Three important foliar diseases of tea - Alternaria blight caused by *Alternaria alternata*, black rot caused by *Corticium theae* and brown blight caused by *Colletotrichum gloeosporioides* are prevalent in Dooars and Terai region of North Bengal. In the present paper polyclonal antibody-based immunoassays have been standardized for early detection of these three foliar fungal pathogens.

Materials and Methods

Plant material : Eighteen tea varieties - seven from Tocklai (TV-18, TV-22, TV-25, TV-26, TV-29, TV-30, T-17/1/54), six from UPASI(UP-2, UP-3, UP-9, UP- 26, BSS-2, BSS-3) and five from Darjeeling (CP- 1/1, BS/7A/76, P-312, AV-2, TS-449) maintained in Tea Germplasm Bank, Department of Botany, University of North Bengal, were used for experimental purpose.

Fungal cultures : *Colletotrichum gloeosporioides* [telomorph *Glomerella cingulata* (Stoneman) Spauld and Schrenk] was isolated from naturally infected tea plants (TV-22) in Phytopathological Experimental Garden and subsequently identified (W7659) from the Diagnostic and Advisory Service, CABI Bioscience UK Center. *Alternaria alternata* (W8053), a foliar fungal pathogen was isolated from young tea leaves of T-17 and also identified by Gobal Plant Clinic, Diagnostic and Advisory services, CABI Bioscience Centre, UK and *Corticium theae* was obtained from Tocklai Research Station, Jorhat, Assam. The cultures were maintained on PDA slants.

Inoculation technique and disease assessment : Detached leaf inoculation technique as described by Dickens and Cook (1989) was followed for artificial inoculation of leaves. Percent drops that resulted in lesion production were calculated after 48 , 72 and 96 hours of inoculation as described by Chakraborty and Saha (1994), diameter of the lesions was noted. Observations were made based on 50 inoculated leaves for each treatment in average of three separate experiments.

Preparation of antigens : Antigens were prepared from healthy and artificially inoculated tea leaves and mycelia of three foliar pathogens (*C. gloeosporioides, A. alternata* and *C. theae*) following the method of Chakraborty and Saha (1994). Protein content of tea leaf and mycelial antigens was estimated following the method as described by Lowry *et al* (1951). Quantity of protein was estimated from the standard curve made with bovine serum albumin (BSA) as standard.

Production and purification of polyclonal antibody : New Zealand white male rabbits were used to raise polyclonal antibodies against mycelia antigens of *Alternaria alternata, Colletotrichum gloeosporioides* and *Corticium theae* separately following the method of Chakaraboprty and Purakayastha (1983). Normal sera were collected by marginal ear vein puncture before immunization. The antigen emulsified with equal volume of Freund's complete/incomplete adjuvant were injected subcutaneously at weekly intervals for six consecutive weeks. The blood samples were collected after six weeks following injection and were kept for 1h at 30°C. The clots were loosened and stored at 4°C. The antisera were then clarified by centrifugation and stored at -20°C until required. Following the protocol of Clausen (1988), IgGs were purified by DEAE-Sephadex column chromatography.

Immunodiffusion : Agar gel double diffusion tests were perfomed using PAb raised against the three pathogens separately following the method of Ouchterlony (1967).

Plate trapped antigen coated enzyme linked immunosorbent assay (PTA-ELISA) : PTA-ELISA was performed essentially as described by Chakraborty *et al* (1996). Antigens from host leaf and fungal pathogens were diluted with coating buffer and IgGs were diluted 1: 125 with PBS-Tween containing 0.5% BSA. Goat antirabbit IgG (whole molecule) alkaline phosphatase (Sigma) conjugate and 4-nitrophenyl phosphate (pNPP) as enzyme-substrate were used for ELISA tests. Absorbance values were measured at 405nm in an ELISA reader (Multiskan EX, Labsystems). Absorbance values in wells not coated with antigens were considered as blanks.

Dot immunobinding assay (DIBA) : Antigens were prepared from healthy and *A.alternata*, *C. gloeosporioides* and *C. theae* inoculated tea leaves separately. These were then loaded on nitrocellulose membrane filters using Bio-Dot apparatus (Bio-Rad). Dot immunobinding assay was performed using PAbs of *A.alternata*, *C. gloeosporioides* and *C. theae* as outlined by Lange *et al* (1989).

Indirect Immunofluorescence : Immunofluorescence of the mycelia and the spores was done as described by Chakraborty *et al* (1995). Mycelia and spores were washed with phosphate buffered saline solution (PBS) at pH 7.5 by centrifugation at low speed, treated with normal sera or antisera diluted (1: 50) in PBS and incubated for 1h at room temp. These were washed thrice with PBST as mentioned above and treated with goat anti-rabbit immunoglobulinG (IgG) whole molecule conjugated with fluorescein isothiocyanate (Sigma) diluted 1 : 40 with PBS (pH 7.2) and incubated in dark for 45 min at room temp, again washed in PBS and mounted in 10% glycerol. The observations were made under UV light using a Leica Leitz Biomed microscope with fluorescence optics suitably equipped with 13 filter. Photographs were taken using 800 ASA film.

Results

Initially screening of resistance of tea varieties was performed against three foliar fungal pathogens (*C. gloeosporioides*, *A. alternata* and *C. theae*) following detached leaf inoculation technique. Results revealed that among eighteen tea varieties tested against brown blight pathogen (*C. gloeosporioides*) six varieties (TV-22, T-17/1/54, BSS-3, UP-26, CP-1/1 and AV-2) were found to be highly susceptible, while only one variety (UP- 3) was moderately susceptible, whereas two varieties (TV-30 and BS/7A/76) were highly resistant, and the remaining nine varieties were moderately resistant. Artificial inoculation of detached leaves of 9 Tocklai varieties and 6 UPASI varieties were carried out with *C. theae*. TV-20, TV-25, TV-26 and BSS-3 were found to be highly resistant, whereas TV-9, TV-29 and UP-26 were moderately resistant. The remaining varieties (TV-18, TV-22, TV-23, UP-2 and UP-3) were highly susceptible, while T-17/1/54, UP-9 and BSS-2 were found to be moderately susceptible.

Effectiveness of PAbs separately raised against the three foliar fungal pathogens (*C. gloeosporioides*, *A. alternata* and *C. theae*) of tea were initially checked by agar gel double diffusion test. Optimization of PTA- ELISA was done considering two variables - dilution of the antigen extract and dilution of the antiserum to obtain maximum sensitivity. Alkaline phosphatase dilution was kept at 1:10,000 in all the tests. Doubling dilutions of mycelial antigens were tested against two antiserum dilutions (1:125 and 1: 250). ELISA values increased with increasing concentration of antigen. In case of *C. gloeosporioides* antiserum absorbance value of 0.9 were obtained in case of 1:125 dilution for antigen concentration as low as 81.25 ng/ml. Higher values were obtained for 1:125 dilution than 1: 250. The absorbance values greater than 3 were obtained when the saturation was reached. After that the accuracy reduced, as the absorbance decreased on increasing concentration of antigen. In case of *C. theae*, antigen dilutions of *C. theae* ranging from 10,000 μg/ml to 78 μg/ml were tested against two antisera dilutions (1:125 and 1:250). ELISA values increased with the concomitant increase of antigen concentration. Mycelial antigen concentrations as low as 62.5 μg/ml could also be easily detected by ELISA at both antisera dilutions (Fig 1 A). PTA-ELISA reactions with different dilutions of three different antibodies ranging from 1:16,000 to 1:125 were determined separately at antigen concentration 62.5 μg/ml. Absorbance values decreased with increasing dilution (Fig. 1 B) . Absorbance values in ELISA decreased from the dilution of 1:125 to 1: 16,000. Highest absorbance values were obtained with 1:125 antiserum dilution in all the cases. Thus, for optimum results, 1:125 dilution of antibody was used in all the immunological experiments.

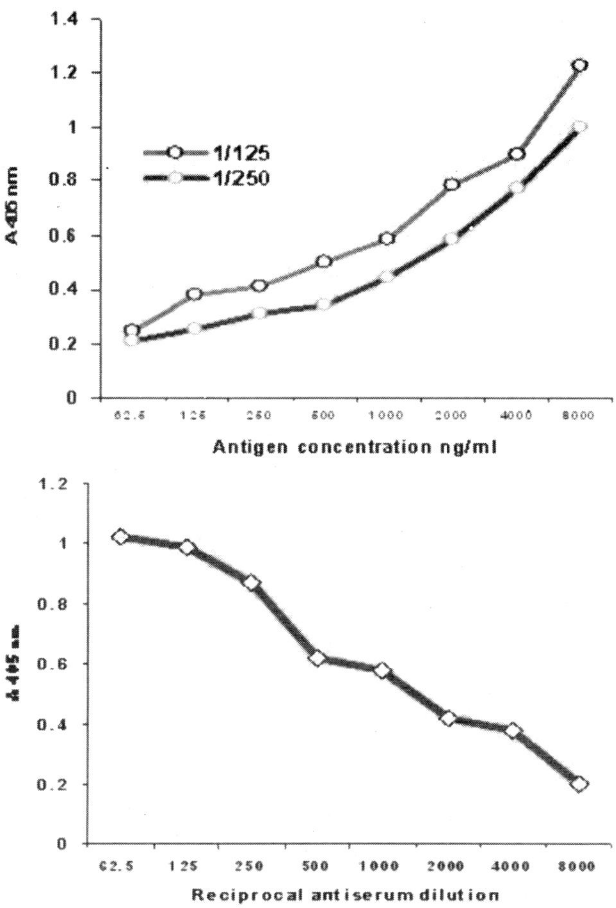

Figure 1 :
A. Optimizaton of antigen concentrations of *A. alternata* with homologous PAbs using PTA-ELISA format.
B. Optimization of PAb raised against mycelia of *A. alternata* with homologous antigen

For brown blight pathogen, the antigen concentration was adjusted to 60 g/ml. The results of immunological screening by PTA-ELISA are presented in Table 1.

Table 1 : PTA-ELISA reaction values of healthy and *C. gloeosporioides* inoculated leaves of tea using PAb of *C. gloeosporioides*

Antigens (40 mg/ml) Varieties	Healthy	Absorbance values at A_{405} Inoculated
TV 22	0.699 ± 0.051	1.250 ± 0.051
T-17/1/54	0.659 ± 0.031	1.506 ± 0.030
BSS -3	0.627 ± 0.063	1.280 ± 0.041
UP -26	0.502 ± 0.055	0.954 ± 0.058
CP - 1/1	0.579 ± 0.058	1.026 ± 0.0 36
AV -2	0.562 ± 0.039	1.013 ± 0.038
TV 30	0.356 ± 0.039	0.859 ± 0.035
BS/7A/76	0.329 ±0.062	0.828 ± 0.044
UP -3	0.404 ± 0.061	0.902 ± 0.056

These values were correlated with the pathogenicity test values in case of infected leaves. There was a positive correlation coefficient (r= 0.986), which was significant at P=0.05 level. The varieties showing greater susceptibility had higher PTA-ELISA values, while those with greater resistance showed lower absorbance values. For *C. theae*, tea leaf antigens were prepared from 6 UPASI varieties (UPASI-2, UPASI-3, UPASI-9, UPASI-26, BSS-2, BSS-3) and 9 Tocklai varieties (TV-9, TV-18, TV-20, TV-22, TV-23, TV-25, TV-26, TV-29 and TeenAli-17/1/54), leaf antigens of non-host (*Oryza sativa*), mycelial and cell wall antigens two isolates of *C.theae* and non-pathogen (*Fusarium oxysporum*). All of these antigens at a concentration of 40 μg/ml were tested against 1:250 dilution of PAb raised against *C.theae* by using PTA-ELISA formats. The susceptible varieties had greater ELISA values than the resistant varieties. Inoculated samples always showed higher ELISA values than the healthy ones (Table 2).

PAb raised against mycelia of *A. alternata* was tested against 14 tea varieties following artificial inoculation with test pathogen. Healthy and artificially inoculated leaves were harvested after 70h for extraction. Results (Fig 2) revealed that all tested varieties gave higher absorbance values in PTA-ELISA format in inoculated leaf antigens than the healthy but the difference was significant in susceptible varieties.

Table 2 : Absorbance values in PTA-ELISA for reactions of PAb of *C. theae* with healthy and artificially inoculated tea leaf antigen

Antigens (40 mg/ml) Varieties	Healthy	Absorbance values at A_{405} Inoculated
TV -18	1.423 ± 0.12	1.889 ± 0.17
TV -22	1.434 ± 0.22	1.763 ± 0.07
UAPSI -2	1.348 ± 0.42	1.712 ± 0.09
BSS -2	1.103 ± 0.22	1.648 ± 0.14
TV -23	1.723 ± 0.52	1.973 ± 0.12

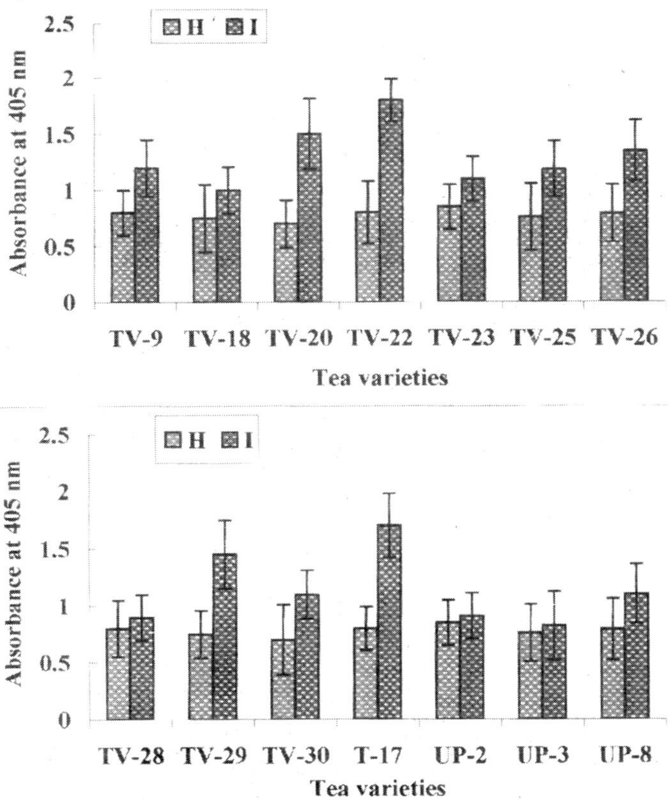

Figure 2 : ELISA response of PAb of *A. alternata* with antigens of healthy and inoculated tea leaf tissues

In dot immunobinding assay greater colour intensity was noticed in TV-22 and T-17 varieties which showed susceptible reactions in varietal resistant test. TV-30 and BS/7A/76, on the other hand showed insignificant colour. These were incompatible with *C. gloeosporioides*. In the present work, three foliar fungal pathogens were detected by molecular methods which is an invaluable tool for the modern detection.

When mycelia of *A. alternata*, *C. gloeosporioides* and *C. theae* were treated with their respective PAbs, bright apple green fluorescence was observed that was intense in young hyphal tips. Indirect fluorescence of the spores of *A. alternata* and *C. gloeosporioides* following treatment with the homologous PAbs showed apple green fluorescence, intense along the spore wall (Fig.3A). Treatment of young germinated sclerotia of *C.theae* with homologous PAb and FITC showed a general fluorescence that was also more intense on young hyphal tips (Fig.3B). Healthy leaf tissue indicated autofluorecence on the cuticle (Fig.3C), while infected leaf tissue showed bright fluorescence on mesophyll tissues (Fig. 3D).

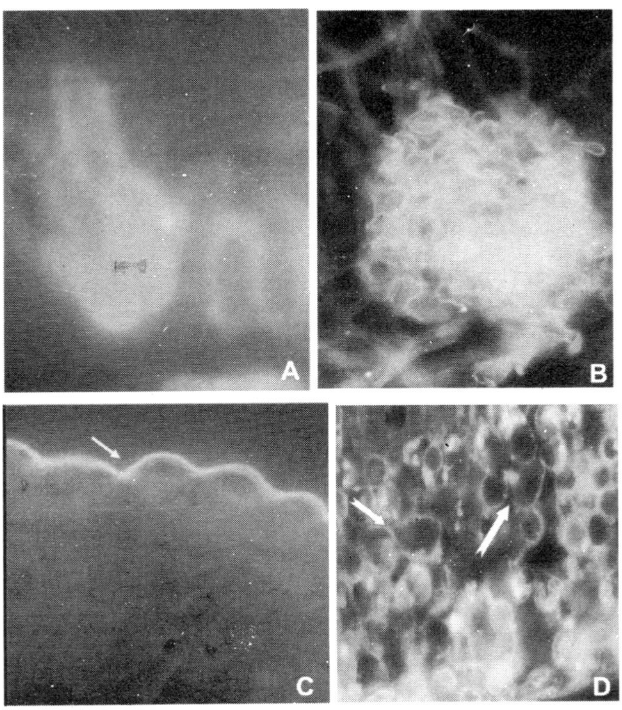

Figure 3 : Indirect immunofluorescence of conidia of *C. gloeosporioides* (A) and sclerotium of *C. theae* (B), Cross section of tea leaf, healthy (C) showing auto-fluoroscence and infected (D) showing bright fluorescence on mesophyll tissues.

Discussion

The results of the present findings indicate that there is little cross-reactivity among the three foliar fungal pathogens - *A.alternata*, *C. gloeosporioides* and *C. theae*. The polyclonal antibody raised against these pathogens can be successfully used to detect each the three diseases separately - Alternaria blight, brown blight and black rot. Besides, tea germplasm can also be screened to find out the most resistant varieties by ELISA and Dot-blot, which are fast and dependable as compared to the traditional methods. Besides, early detection of pathogens by ELISA have been reported earlier in tea (Chakraborty *et al.*, 1996), sugarcane (Viswanathan *et al.*, 2000) and ginger (Ghosh and Purakayastha, 2003). In these studies also, just as in the present findings, mycelium of the pathogen was used as antigen to raise the PAbs for detection.

A direct relationship was found between ELISA values and the percentage lesion formation after inoculation with the pathogens in the inoculated tissues. Similarly, the relationship between ELISA values and the pathogen concentration per gram soil was linear in case of *Spongospora subterranea* (Welsh *et al.*, 1996). Besides, polyclonal antibodies are easier to produce than monoclonal antibodies and have been proven to be more sensitive of the two. For instance, Kratka *et al.* (2002) had raise both polyclonal and monoclonal antibodies to detect *Colletotrichum acutatum* by PTA-ELISA. According to their findings only the polyclonal antiobody was sensitive enough to recognize the pathogen. Also, as suggested by Ghosh *et al.* (2005), ELISA is the more sensitive method, but the field applicability of dot blot is much greater. The present findings support this view in case of all the three foliar pathogens of tea. Similarly, indirect ELISA has also proved to be an invaluable tool for screening commercially cultivated tea for their susceptibility to *Exobasidium vexans*, another foliar fungal pathogen of tea (Chakraborty and Sharma, 2007).

Therefore, the present findings indicate that the antibodies raised against the mycelia of foliar fungal pathogens of tea can be easily packaged into PTA-ELISA formats for quick and efficient screening of tea varieties and detection of the disease prior to symptom development for developing disease management strategies. Dot blot can be used as an alternative to be used in the field.

References

Chakraborty B N Basu P Das R Saha A and Chakraborty U. (1995). Detection of cross-reactive antigens between *Pestalotiopsis theae* and tea leaves and their cellular location. *Ann. Appl.Biol.* 127: 11-21.

Chakraborty BN and Chakraborty U. (2003). Immunodetection of plant pathogenic fungi. In : GP Rao, C. Manoharachari, DJ Bhatt, RC Rajak and TN Lkhanpal (Eds) *Frontiers of fungal diversity in India*. International Book distributing company, Lucknow. Pp 23-41.

Chakraborty B N and Chakraborty U. (1998). Tea and Human Health In : *Phytodiversification and Human Welfare* (Chapter18) (Eds. S.K.Varma, J.V.V.Dogra and A.K.Roy) M.D.Publications, New Delhi. p.197-207.

Chakraborty BN and Purakayastha RP. (1983). Serological relationship between *Macrophomina phaseolina* and soyabean cultivars. *Physiol Pl Pathol* 23 : 197-205

Chakraborty BN and Saha A. (1994). Detection and cellular location of cross-reactive antigens shared by *Camellia sinensis* and *Bipolaris carbonum*. *Physiol. Mol. Plant Pathol.* 44: 403 - 416.

Chakraborty BN and Sharma M. (2007). Serological detection and immunogold localization of cross-reactive antigens shared by *Camellia sinensis* and *Exobasidium vexans*. *J Appl Microbiol* 103 (5) : 1669-1680.

Chakraborty U Basu P Das R Saha A and Chakraborty B N. (1996). Evaluation of antiserum raised against *Pestalotiopsis theae* for the detection of grey blight of tea by ELISA. *Folia Microbiol.* 41: 413 -418.

Clausen J. (1988). Immunochemical techniques for the identification and estimation of macromolecules. In:*Laboratory techniques in Biochemistry and Molecular Biology*, Burden, R.H. and Van Kinppenberg, P.H., (Eds.) Elsevier.The Netherlands. 1(3): 464-65.

Dickens J S and Cook RTA. (1989). *Glomerella cingulata* on *Camellia*. *Plant Pathol.* 38: 147-168.

Gawande SJ, Chimote VP, Shukla A and Gadewar AV (2006). Serological and molecular diagnosis of plant pathogen In : Gadewar AV and Singh BP (eds.) *Plant protection in NewMillennium*. New Delhi. Satish Serial Publishing House, pp 569-588.

Ghosh R and Purakayastha RP. (2003). Molecular diagnosis and induced systemic protection against rhizome rot disease of ginger caused by *Pythium aphanidermatum*. *Curr Sci* 85: 1782-1783.

Ghosh S, Rawat P, Gupta SC and Singh BP. (2005). Comparative diagnostic potentiality of ELISA and Dot-ELISA in prepatent diagnosis of experimental *Fascicola gigantica* infection of cattle. *Indain J Exp Biol* 43: 536-541.

Hajra GN. (2001) .Tea Cultivation. Comprehensive Treatise. -International Book Distributing Company, Lucknow.

Jayaraman J. (1996). Laboratory Manual in Biochemistry, New Age Publishers

Kennedy R, Wakeham AJ, Byrne KG, Meyer UM and Dewey FMA. (2000). New method to monitor airborne inoculums of the fungal plant pathogens *Mycosphaerella brassicicola* and *Botrytis cinerea*. *Appl. Environ Microbiol* 66 : 2996-3003.

Kratka J, Pekarova KRB, Kudlikova I, Slova EJand Zemankova M. (2002). Utilization of immunocytochemical methods for detection of *Colletotrichum* spp. In strawberry. *Pl Prot Sci* 38 : 55-63.

Muraleedharan N Chen ZM. (1997). Pests and diseases of tea and their management. - *J. Plantation Crops* 25: 15-43.

Ouchterlony O .(1976). Immunodiffusion and immunoelectrophoresis. In : D. M. Weir (Ed.) *Handbook of Experimental Immunology*. Blackwell Scientific Publishers, Oxford and Edinburgh, p. 655.

Viswanathan R, Padmanaban P, Mohanraj D and Jothi R. (2000). Indirect- ELISA technique for the detection of red rot pathogen in sugarcane (*Saccharum* spp. hybrid) and resistance screening. *Indian J Agri Sci* 70 : 308-311.

Welsh JA, Merz U and Harrison JG. (1996). Serological detection in soil of *Spongospora subterranea* and quantification in soil. *Pl Pathol* 45 : 884-893.

❑❑❑

Chapter 5

DUAL APPLICATION OF *GLOMUS MOSSEAE* AND *BACILLUS PUMILUS* IN ENHANCING GROWTH OF TEA AND SUPPRESSION OF SCLEROTIAL BLIGHT DISEASE

B.N. CHAKRABORTY, U. CHAKRABORTY, U. DE, A.P. CHAKRABORTY AND K. RAI
Immuno-Phytopathology Laboratory, Department of Botany
University of North Bengal, Siliguri- 734 013, West Bengal, India

Introduction

Rhizosphere is the habitat in which several biologically important processes and interactions takes place and the microorganisms that grow in the rhizosphere are ideal for use either as biological control agents or plant growth promoters. The use of beneficial microorganisms as biofertilizers and biocontrol agents has assumed importance in recent years, because these are environmentally safe and do not cause pollution (Mathivanan *et al.* 2005). Rhizosphere soil of plants contain a wide variety of microorganisms- bacteria, fungi and actinomycetes, some of which are beneficial, while others are harmful. Among the beneficial microorganisms, two groups are very important agriculturally – the mycorrhizal fungi and plant growth promoting bacteria (PGPR). Mycorrhizal fungi colonize plant roots and extend the root system into the surrounding soil. Estimates of amounts of mycorrhizal filaments present in healthy soil are astonishing. Several miles of filaments can be present in less than a thimbleful of soil associated with vigorously growing plants. The relationship is beneficial because the plant enjoys improved nutrient and water uptake, disease resistance and superior survival and growth.

Tea (*Camellia sinensis* (L.) O. Kuntze) is an important plantation crop grown in tropical agro climate. India is one of the major producers, consumers and exporters of tea. Being a perennial, it is susceptible to attacks by fungal and insect pests. Extensive use of fungicides and pesticides in tea plantations, combined with over doses of chemical fertilizers has caused the problem of pesticide residues in tea leaves, and has also deteriorated the condition of soil in several cases, leading to decrease in productivity. However, growing concerns over presence of chemical residues in tea leaves has intensified the search for alternatives. Shift in priorities has resulted in major thrust being given to biological control as components of integrated disease management. Several mechanisms have been postulated to explain how the plant growth promoting rhizobacteria stimulate plant growth, which may be either direct or indirect (Kloepper 1993). However, it is now evident, that in field conditions, the above traits alone may not be sufficient to account for the observed growth promotion. The biochemical or physiological changes induced in host, that are activated by the PGPRs, also lead to plant growth promotion and resistance in the host against specific pathogens (Chakraborty *et al.* 2006). The mechanism by which PGPRs mediate ISR are, to a large extent unknown. None of the mechanisms of action are mutually exclusive, and frequently, a single agent exhibits several modes of action. It has been reported that many soil bacteria can solubilize inorganic phosphates which can play an important role in supplementing phosphorus to the plants, allowing a sustainable use of phosphate fertilizers.

The dual activity of AMF and PGPR is generating ample evidence to act as bioprotector as well as biofertilizer which play a significant role in sustainable role in agriculture.

In the present study, a strain of PGPR- *Bacillus pumilus*, and a mycorrhizal fungus- *Glomus mosseae*, were selected and tested for their effects on growth promotion and disease suppression of tea plants. Mechanism of the observed effects by the dual application of the microorganisms were determined both biochemically and immunologically.

Materials and Methods

Test organisms *:. Bacillus pumilus* TRS 3 was isolated from the rhizosphere soil of tea bushes from Hansqua Tea Estate, West Bengal, India, and preliminarily identified on the basis of morphological, microscopic and biochemical tests. *Sclerotium rolfsii* was obtained from the culture collection of Immuno-Phytopathology Laboratory,

Department of Botany, University of North Bengal, and was maintained by regular sub-culturing in PDA.

***Isolation of Arbuscular mycorrhizal fungi*.:** Spores of arbuscular mycorrhizal fungi were isolated from rhizosphere soil of tea by wet sieving and decanting method (Gerdemann and Nicolson, 1963). Tea roots were inoculated with *G. mosseae* alone and in combination with *B. pumilus* which was applied as soil drench.

Tea Plants : Five different varieties of tea (TV-29, TV-25, TV-23, TV-26 and UP-2) were selected for experimental purposes.

Application of bacteria and fungus to soil : Initially *B. pumilus* was cultured in Nutrient Broth medium and was allowed to grow properly, with shaking at 37°C at 120 r.p.m for 48 h. At the end of the log phase, bacterial cells were collected by centrifugation and aqueous suspensions were made and diluted as necessary to maintain the bacterial concentration at 1.28×10^9c.f.u/ml. The aqueous suspension was then applied as a soil drench, @ 50 ml/plant to the rhizosphere of 5 varieties of tea seedlings in potted conditions one month after transplantation. Application was done twice at an interval of one month.

Culture of *S. rolfsii* was grown in sand- maize meal medium (maize meal: sand: water- 1:9:1.5 w:w:v) in autoclavable plastic bags (sterilized at 20 lbs. pressure for 20 min) for a period of three weeks at 28°C until the mycelia completely covered the substrate.

G. mosseae was inoculated with the spores at an average of 40-45 spores in each pot. In case of combined inoculation of AMF and *B. pumilus*, spores of *G. mosseae* were inoculated prior to the application of *B. pumilus*.

Growth of Tea seedlings :Growth promotion was studied in terms of increase in height, number of leaves and biomass in potted plants. Observations were recorded after 1 and 2 months of application. In each treatment, 3 replicates were taken and average of the 3 replicate plants were analysed.

Extraction and quantification of soil phosphate :Soil sample (1g) was air dried and suspended in 25 ml of the extracting solution (0.025N H_2SO_4, 0.05N HCl) to which activated charcoal (0.01g) was also added, shaken well for 30 min on a rotary shaker and filtered through Whatman No. 2 filter paper (Mehlich 1953). Quantitative estimation of phosphate was done following ammonium molybdate-ascorbic acid method as described by Knudsen and Beegle (1988).

Enzyme assays : β-1,3- glucanase (EC.3.2.1.39):Extraction and assay of enzyme activity of β-1,3- glucanase was done following the method described by Pan *et al.* (1991). Chitinase (EC. 3.2.1.39): Chitinase (EC. 3.2.1.39) was extracted and assayed by following the method described by Boller and Mauch (1988) with modifications. Phenylalanine ammonia lyase (EC.4.3.1.5): PAL was extracted from leaves and assayed following the method described by Chakraborty *et al.* (1993) with modifications. Peroxidase: Peroxidase was extracted and assayed following Chakraborty *et al.* (1993).

In vitro antagonistic tests :*In vitro* antagonism of *B. pumilus* to fungal pathogens was determined by paired culture on nutrient agar medium as described by Chakraborty *et al.* (2006).

Inoculation of pathogen and disease assessment : Tea sedlings were inoculated by adding 100g of previously prepared inoculum of *S. rolfsii* to the rhizosphere soil. Inoculation was done 3 days after final application of *B. pumilus* and *G. mosseae*, singly or jointly. Disease assessment was performed following the method of Chakraborty *et al.* (2006) after 15, 30 and 45 days of inoculation.

Production and purification of polyclonal antibodies : Polyclonal antibodies were raised against chitinase in white, male rabbit following the procedure described by Alba and Devay (1985).

Immunological assays

PTA-ELISA Plate trapped antigen ELISA was performed following the method as described by Chakraborty *et al.*(1996).

DOT-BLOT Dot immunobinding assay was carried out using the antigen and PAb raised against chitinase following the procedure suggested by Lange *et al.* (1989).

Results and Discussion

Among the AM fungi, *Glomus mosseae* could be determined as the predominant one, followed by other genera such as *Gigaspora*,

Acaulospora and *Scutellospora* (Fig.1 A-L). On the basis of consistent association of *Glomus mosseae*, it was selected for further experiments. *B. pumilus* was tested for different PGPR activities (Fig. 2 A-F). Results revealed that the bacterium could solubilize phosphate, produce siderophore, volatiles as well IAA (Table 1). In paired culture, *B. pumilus* could inhibit the growth of *Sclerotium rolfsii* (Table 2). Application of *B.pumilus* and *G.mosseae* in the rhizosphere of potted tea plants led to an increase in the growth of seedlings in terms of increase in height and number of leaves (Tables 3 & 4). Rhizosphere of tea plants was inoculated by *B.pumilus* and *G.mosseae* prior to challenge inoculation with *Sclerotium rolfsii*. Development of root rot was determined after 15, 30 and 45 days of inoculation on the basis of root rot index. Results revealed that both microorganisms could reduce root rot, but maximum suppression of disease was due to joint inoculation (Table 5).Mathivanan *et al.* (2005) also obtained synergistic effect of *Pseudomonas fluorescens* and *Trichoderma viride* in plant growth promotion, yield enhancement and disease suppression in rice.

Table 1 : *In vitro* PGPR characteristics of *B. pumilus*

Characteristics	***B. pumilus***
Phosphate solubilisation	+
Siderophore production	+
Protease production	+
Chitinase production	-
HCN production	-
Volatile production	+
IAA production	+

+= Activity present ; −= Activity absent

Total phosphate content of soil was determined after application of the microorganisms. Results revealed that soil P content had decreased due to application indicating that the plant could uptake phosphorous which had been solubilized by the microorganisms (Table 6). Phosphorus is one of the major plant nutrients. Its assimilation, storage and metabolism are of major importance to plant

Figure 1. A. *Glomus mossea* ; B. *Glomus mosseae* ; C. *Glomus drummondii* ; D and E. *Glomus badium* ; F and G. Extraradical hyphae ; H. *Glomus fasciculatum* ; I. *Glomus* sp., ; J. Intraradical hyphae ; K-L. Arbuscules and Vesicles

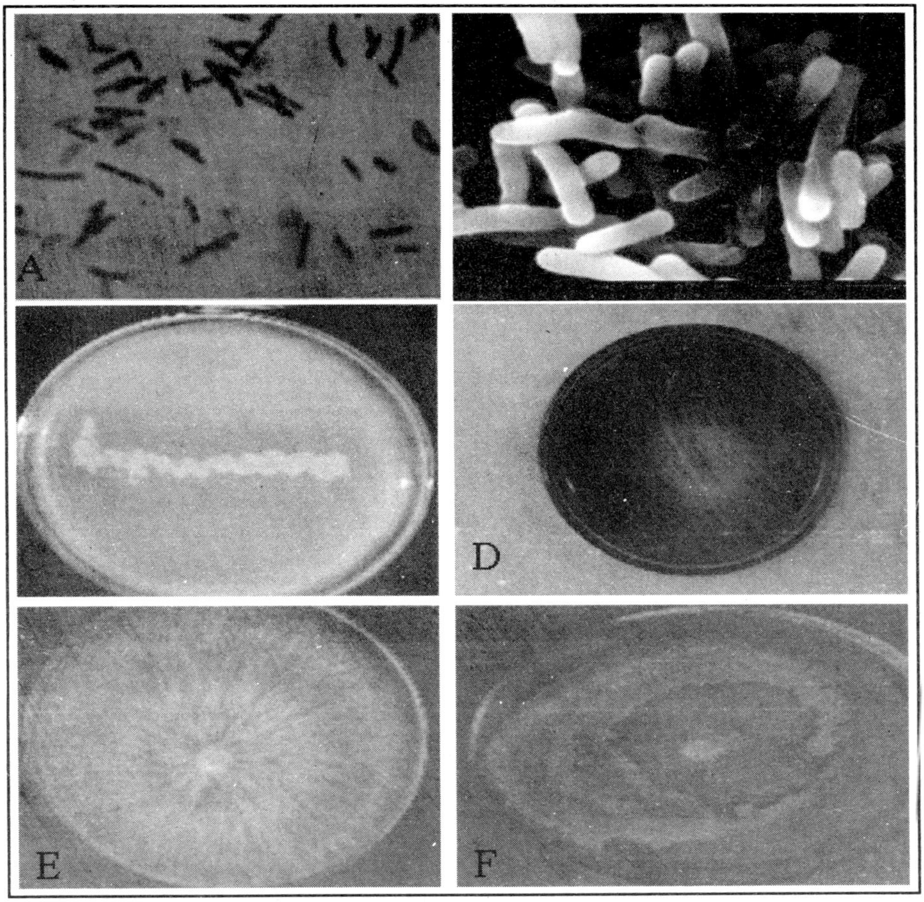

Figure 2 : Light and Scanning electron microscopic view of *Bacillus pumilus* (A&B), Phosphate solubilisation in PVK medium (C), Siderophore production in CAS medium (D), Antagonistic activity of *B. pumilus* against *Sclerotium rolfsii* (F), (E)- respective control.

Table 2 : *In vitro* antagonistic tests of *B.pumilus* against *Sclerotium rolfsii*

Test fungi	Paired with	Dia. of fungal growth (cm) after 7 days	Zone of inhibition (cm)	% of inhibition	Mycelial dry wt.(mg)
S. rolfsii	Nil	8.8±1.3	-	-	465
	B. pumilus	1.3 ±0.29	2.2 ±0.22	86 .09±2.66	190

Average of 3 replicates; ±=S.E.

growth and development (Duff *et al.* 1994). However, a greater part of soil phosphorus, approximately 95–99% is present in the form of insoluble phosphates and hence cannot be utilized by the plants. To increase the availability of phosphorus for plants, large amounts of fertilizer are used on a regular basis.

Table 3 : Growth promotion in tea varieties following the application of AMF

Tea Varieties	Treatment	% increase in height 1 month after treatm ent	2 months after treatment	% increase in leaf no. 1 month after treatment	2 months after treatment
TV -29	Control	8.69	9.2	9.0	10 .0
	Glomus mosseae	14 .0	15.1	13.33	14 .0
TV -23	Control	8.0	9.1	18.18	19.2
	Glomus mosseae	35.8	36 .0	72.2	73 .0
TV -25	Control	15.38	17 .0	46.15	46.2
	Glomus mosseae	25.80	25.81	47.61	48 .0
TV -26	Control	15 .0	15.5	15.38	16 .0
	Glomus mosseae	44.82	45 .0	38.8	38.8
UP -2	Control	26.6 0	27 .0	20 .0	21 .0
	Glomus mosseae	32.35	35 .0	36.84	37.2

Table 4 : Growth promotion in tea varieties following the application of PGPR

Tea Varieties	Treatment	% increase in height 1 month after treatm ent	2 months after treatment	% increase in leaf no. 1 month after treatment	2 months after treatment
TV -29	Control	17.72	18	4.16	5.2
	Bacillus pumilus	20	21	24.13	25.1
TV -23	Control	7.14	9.2	22.2	23
	Bacillus pumilus	21.56	24.2	26.08	27
TV -25	Control	3.9	4.0	19.04	21
	Bacillus pumilus	20.58	21.1	44.4	44.5
TV -26	Control	6.3	7.0	4.76	4.8
	Bacillus pumilus	10.5	11	27.27	28.2
UP -2	Control	8.3	12	5.26	5.3
	Bacillus pumilus	15.15	18	38.7	39

Table 5 : Sclerotial blight development in tea roots in presence and absence of *G. mosseae* and *B. pumilus* 10 pots/treatment; *S.r.= Sclerotium rolfsii; G.m= Glomus mosseae; B.p=Bacillus pumilus.*

Tea varieties	Disease index (45 days after inoculation)			
	S.r.	***S.r.+G.m***	***S.r.+ B.p***	***S.r.+ G.m+ B.p***
TV -29	4.1	3.0	1.9	1.3
TV -23	5.9	3.9	2.2	1.2
TV -25	3.5	2.8	1.7	1.1
TV -26	4.0	2.4	2.1	2.0
UP -2	4.0	2.7	2.1	1.6

Disease (Sclerotial Blight) Index computed on a scale of 0-6 on the basis of underground and above ground symptoms. Rot index: 0- no symptoms; 1- small roots turn brownish and start rotting; 2- leaves start withering and 20-30% of roots turn brown; 3- leaves withered and 50% of roots affected; 4- shoot tips also start withering; 60-70% roots affected; 5- shoots withered with defoliation of lower withered leaves, 80% roots affected; 6-whole plants die, with upper withered leaves still remaining attached; roots fully rotted.

Table 6 : Phosphate estimation from control and treated potted soil

Tea varieties	Treatment	Soil phosphate (µg/g tissue)
TV -29	Control	39.64
	B.pumilus	29.02
	G. mosseae	28.42
TV -23	Control	92.76
	B.pumilus	72.11
	G. mosseae	72.41
TV -25	Control	167.1
	B.pumilus	43.39
	G. mosseae	32.91
TV -26	Control	26.18
	B.pumilus	18.55
	G. mosseae	19.45
UP -2	Control	46.38
	B.pumilus	31.42
	G. mosseae	32.4

Activities of defense enzymes and phenolics (Tables 7&8) were assayed in leaves of tea seedlings subjected to various treatments. Activities of all enzymes were enhanced due to the various treatments in all varieties but comparatively better responses were observed in TV-26 by dual application of *G. mosseae* and *B. pumilus* (Figs. 3A,B,C). Induction of defense related enzymes by *P. fluorescens* in black pepper and *Phytophthora capsici* pathosystem was reported by Paul and Sharma (2003). Two isolates of *B. pumilus* were reported to be best plant growth promoters and biocontrol agents against downy mildew disease in pearl millet (Niranjana *et al.*, 2003).

Table 7 : Phenolic contents in tea leaf following treatments with AMF & PGPR

Tea Varieties	Treatment	Total Phenol (mg/g tissue)	O -dihydroxy phenol (mg/g tissue)
TV -29	Control	25	08
	Glomus mosseae	30	10
	B.pumilus	35	12
	Glomus mosseae+ B.pumilus	36	14
TV -23	Control	15	07
	Glomus mosseae	24	11.5
	B.pumilus	32	11
	Glomus mosseae+ B.pumilus	28	12.6
TV -25	Control	32	5.5
	Glomus mosseae	45	8.2
	B.pumilus	46	9.0
	Glomus mosseae+ B.pumilus	41	8.8
TV -26	Control	30	6.0
	Glomus mosseae	32	12
	B.pumilus	38	10.5
	Glomus mosseae+ B.pumilus	38	12.7
UP - 2	Control	22	.09
	Glomus mosseae	36	10
	B.pumilus	35	15
	Glomus mosseae+ B.pumilus	39	15.5

Average of 3 replicates

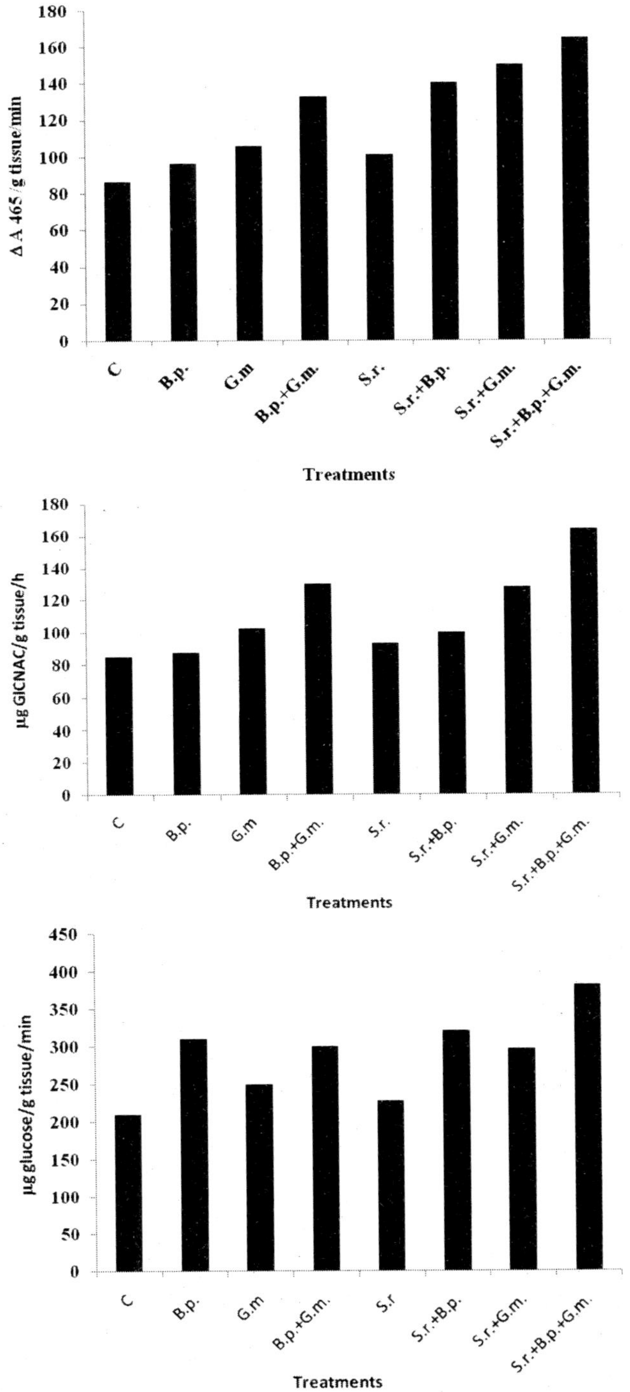

A

B

C

Figure 3 A,B,C Activities of peroxidase (A), chitinase (B) and b-1,3 glucanase (C) in leaves of tea plant (TV-26) following application of *B.pumilus* and *G.mosseae*.

Table 8 : Effect of application of AMF and PGPR to potted soil on enzyme activity in tea

Tea Varieties	**Treatment**	**Peroxidase (Δ O.D. / gm tissue/min)**	**PAL(μgm cinnamic acid/gm tissue)**
TV-29	Control	1.5	85
	Glomus mosseae	2.1	92
	Bacillus pumilus	2.5	90
	G. mosseae+ B.pumilus	**2.7**	**91**
TV-23	Control	4.2	55
	Glomus mosseae	5.3	71
	Bacillus pumilus	5.5	65
	G. mosseae + B.pumilus	**6.0**	**70**
TV-25	Control	3.3	60
	Glomus mosseae	5.4	81
	Bacillus pumilus	5.0	75
	G. mosseae+ B.pumilus	**5.5**	**71**
TV-26	Control	2.1	45
	Glomus mosseae	2.6	55
	Bacillus pumilus	2.8	51
	G. mosseae+ B.pumilus	**3.1**	**60**
UP-2	Control	4.3	58
	Glomus mosseae	6.0	81
	Bacillus pumilus	5.4	71
	G. mosseae+ B.pumilus	**6.1**	**74**

In order to confirm the induction of enhanced activities of defense enzymes due to treatment with *B.pumilus*, *G.mosseae* or both, immunological tests were done using PAbs raised against enzyme-chitinase. Enzyme extracts were used as antigens and PTA-ELISA and Dot Blot were carried out. Results revealed that ELISA values of reaction of PAbs of chitinase with enzyme extracts from leaves grown in treated soil were higher than the control values. Similarly, in Dot-Blot, more intense dots were observed in treated plants (Table 9).

The overall results of the present study have shown that *B.pumilus* and *G.mosseae* could promote growth and suppress root rot of tea.

While *B.pumilus* acted by both direct and indirect means, *G.mosseae* was responsible mainly for induction of responses within the host. Combined application of both microorganisms gave better results.

Table 9 : ELISA and Dot-Blot values of reactions between PAbs of defense enzymes and enzyme extracts of leaves from treated tea plants

Antigen source*	PAb of chitinase A 405 ELISA	Colour intensity# Dot-Blot
Control	0.306±0.09	-
S. rolfsii inoculated	1.058±0.05	++
S. rolfsii + *B.pumilus* treated	0.006±0.00	-
S. rolfsii + *G.mosseae* treated	0.008±0.00	-
S. rolfsii + *B.pumilus* + *G.mosseae* treated	0.469±0.002	-

* Enzyme extracts from leaves of plants treated as mentioned;
\# Colour Intensity - += Light pink; ++ = Dark pink ; +++ = Deep purplish

Acknowledgement

Financial support from University Grants Commission, New Delhi is gratefully acknowledged.

References

Alba, APC and Devay JE. (1985). Detection of cross- reactive antigens between *Phytophthora infestans*(Mont.) de Bary and *Solanum* species by indirect enzyme- linked immunosorbent assay. *Phytopathology* 112: 97-104.

Boller T and Mauch F. (1988). Colorimetric assay for chitinase. *Meth Enzymol* 161: 430-435.

Chakraborty BN, Chakraborty U, Das R, Basu P and Saha A. (1996). Serological relationship between *Glomerella cingulata* (Stoneman) Spaulid Schrenk and *Camellia sinensis* (L.) O. Kuntze. *J Plant Crop* 24: 205-211.

Chakraborty U, Chakraborty B N. and Basnet M. (2006). Plant growth promotion and induction of resistance in *Camellia sinensis* by *Bacillus megaterium*. *J Basic Microbiol* 46: 186-195.

Chakraborty U, Chakraborty BN and Kapoor M. (1993). Changes in levels of peroxidase and phenylalanine ammonia lyase *Brassica napus* cultivars showing variable resistance to *Leptoshaeria maculans*. *Folia Microbiol* 38: 491-496.

Duff SMG, Sarath G and Plaxton WC. (1994). The role of acid phosphatase in plant phosphorus metabolism. *Physiol Plant* 90: 791-800.

Gerdemann J W and Nicolson T H. (1963). Spores of mycorrhizal *Endogone* species extracted from soil by wet sieving and decanting. *Trans B Mycol Soc* 46: 235-244.

Kloepper JW. (1993). Plant growth promoting rhizobacteria as biological control agents. In Soil microbiol ecology: applications in agricultural and environmental management. (Eds., Metting, F.B. Jr.) Marcel Dekker Inc., NY, USA. pp. 255-274.

Knudsen D and Beegle D. (1988). Recommended phosphorous tests. In: Recommended chemical soil tests procedures for the north central region. WC Dahnke(ed) Bull North Dakota Agric Exp Stn. North Dakota, USA No. 499: 122-15.

Lange L, Heide M and Olson LW. (1989). Serological detection of *Plasmodiophora brassicae* by dot immunobinding and visualization of the serological reaction by scanning electron microscopy. *Phytopathology* 79: 1066-1071.

Lange L, Heide M and Olson LW. (1989). Serological detection of *Plasmodiophora brassicae* by dot immunobinding and visualization of the serological reaction by scanning electron microscopy. *Phytopathology* 79: 1066-1071.

Mathivanan N, Prabhavathy VR and Vijayanandaraj VR. (2005).Application of talc formulations of *Pseudomonas flourescens* Migula and *Trichoderma viride* Pers Ex S.F. Gray decrease the sheath blight disease and enhance the plant growth and yield in rice. *J Phytopathol* 153: 697-701.

Mehlich A. (1953). Determination of P, Ca, Mg, K, Na and NH_4. North Carolina soil test division.

Niranjan S R, Sarosh, B R, Shetty N P, Shetty H S and Reddy M S. (2003). Plausible biochemical and molecular mechanisms involved in plant growth promoting rhizobacterial mediated resistance induction against pearl millet downy mildew disease. In: Proceedings of 6th International Workshop on Plant Growth Promoting Rhizobacteria, 5-10 Oct. 2003. Calicut, India pp. 520-529.

Pan SQ, Ye XS and Kuc J. (1991). A technique for detection of chitinase, b-1,3 glucanase and protein patterns after a single separation using polyacrylamide gel electrophoresis or isoelectric focussing. *Phytopathology* 81: 970-974.

□□□

Chapter 6

ISOLATION AND CHARACTERIZATION OF PLANT GROWTH PROMOTING RHIZOBACTERIA FROM RICE FIELDS OF EASTERN UTTAR PRADESH

S.K. SHAHI[1], A.K. RAI[1], M.B. TYAGI[2], R.P. SINHA[3] AND A. KUMAR[1]

[1] School of Biotechnology, [2] Department of Botany and [3] Centre of Advanced Study in Botany, Banaras Hindu University, Varanasi-221005

Introduction

A plant root is associated with a myriad of both prokaryotic and eukaryotic microorganisms. Typical populations of microorganisms near the root surface are 10^9 bacteria, 10^7 actinomycetes, 10^6 fungi, 10^3 protozoa and 10^3 algae per gram of rhizosphere soil (Ladha and Reddy 2000). Due to fluctuating abiotic and biotic factors including plant metabolites (especially root exudates at different stages of growth and development of plants), wide diversity in microbial population and profile prevails (de Weger *et al.* 1995). In spite of extensive studies made so far, it is still felt that rhizosphere is an ecological niche which is mainly a black box. Such an assumption is based on the fact that several bacteria actively growing in the rhizosphere of a variety of plants are unculturable (de Weger *et al.* 1995, van der Lelie *et al.* 2005). The microbe-plant interaction in the rhizosphere can be beneficial, neutral, variable, or deleterious for plant growth and development. Rhizobacteria that exert beneficial effects on plant development are termed plant growth-promoting rhizobacteria (PGPR). Plant associated nitrogen-fixing bacteria have been considered as one of the possible alternatives for inorganic nitrogen fertilizer for promoting plant growth and yield (Ladha and Reddy 2000). In the last few decades a large array of bacteria including species of *Pseudomonas, Azospirillum, Azotobacter, Klebsiella, Enterobacter, Alcaligens, Arthobacter, Burkholderia, Bacillus* and *Serratia* have been isolated from rhizosphere of various crop plants and reported to enhance plant growth (Kloepper *et al.* 1989; Glick 1995). Interest in the beneficial PGPR has increased recently due to their potential use as biofertilizers (Vessey 2003).

Importance of IAA production by rhizospheric bacteria has been widely acknowledged (Kennedy *et al.* 2004). IAA secreted by bacterium promotes root growth directly by stimulating plant cell elongation or cell division or indirectly by influencing bacterial 1-aminocyclopropane-1-carboxylic acid (ACC) deaminase activity (Pedraza *et al.* 2004). It has been demonstrated that ACC deaminase is involved in the stimulation of root elongation in seedlings. As a matter of fact ACC deaminase hydrolyses plant ACC, the immediate precursor of the phytohormone ethylene and thereby prevents the production of plant growth inhibiting levels of ethylene. Screening of crop plants with extensive root hair formation under field condition may lead to the isolation of novel IAA- over producing bacterium.

Phosphorus is second only to nitrogen as mineral nutrient required by both plants and microbes. In general P in soil immobilize or becomes less soluble either by absorption or chemical precipitation or both. In recent years P solubilizing bacteria have been isolated from different soils (De Los Santos *et al.* 2001). Using radio-labelled insoluble P it has been documented that microbes indeed transfer solubilized P to plants. The level of P solubilization varies in different isolates while growing in the soil. Addition of phosphobacteria to the crop plants shall be an added character in addition to N_2 fixation and IAA production. There is an urgency to isolate novel strains having a combination of composite characters such as N_2 fixation, IAA production and P solubilization.

During the last few years several rhizospheric bacteria endowed with a wide range of beneficial characters such as PGP, MPS, siderophore production and involved in disease control have been isolated (Bottini *et al.* 2004, Miyamoto *et al.* 2004). With the development of newer techniques importantly rRNA sequencing,

RAPD, FISH (fluorescence *in situ* hybridization), DGGE (denaturating gradient gel electrophoresis), SSCP (single-strand-conformation polymorphism) and TGGE (temperature gradient gel electrophoresis), it has now become possible to characterize and reveal genetic diversity among rhizospheric bacteria (Schmalenberger *et al.* 2001). However disclosure of the complex network of interactions which comprises life in the rhizosphere may still reveal various unhidden facts.

To our knowledge little, if any, work has been done on isolation and characterization of PGPR from important crop such as rice from eastern Utter Pradesh (India). This study deals with the isolation and characterization of PGPR from rice rhizosphere of five districts of eastern Utter Pradesh (India).

Materials and Methods

Sample collection and isolation of bacteria: Rhizospheric soil samples from rice fields of different villages falling under the districts of Chandauli, Ghazipur, Jaunpur Mirzapur and Varanasi were randomly collected during the month of November 2007. Soil adhered to the roots of plant of a particular field was carefully removed and mixed together. Ten grams of rhizospheric soil of each field were transferred to a 250 ml Erlenmeyer flask containing 90 ml sterile distilled water and shaken (120 rpm) for 30 min. Serial dilutions (up to 10^{-5}) were made and 0.1 ml aliquots were spread on $JNFb^{-}$ solid agar-agar medium (which contained (in g/l); malic acid (5.0), K_2HPO_4 (0.6), KH_2PO_4 (1.8), $MgSO_4.7H_2O$ (0.2), NaCl (0.1), $Na_2MO_4.2H_2O$ (0.002), Fe-EDTA (4 ml 1.4%) and KOH (4.5), pH was adjusted to 5.8). All the plates were incubated at 37°C in an incubator. Morphologically distinct bacterial colonies appearing after 4-5 days of growth were selected and transferred to fresh plate. Distinct morphotypes of bacteria were characterized on the basis of colony color, pigment formation, elevation, shape and size. Formation of pellicle, if any, was tested after growth in semi-solid medium (0.15% agar-agar) without shaking at 37°C for 96 h.

Estimation of nitrogenase activity : Nitrogenase activity was measured by acetylene reduction assay as per the method of Stewart *et al.* (1967). Equal amount of inoculum from each isolate was added in 3-ml semi-solid (0.15% agar-agar wD v) $JNFb^{-}$ medium in a 7-ml vacutainer tube (Becton-Dickinson, Rutherford, NJ, USA) and grown for 96 h at 30°C. Acetylene was injected in each tube by a hypodermic syringe to attain 10% final concentration and the tubes were incubated at 30°C without shaking. At the desired time interval, 0.2-ml gas phase was removed and the ethylene formed was analyzed in a 5700 Nucon Gas Chromatograph (Nucon Engineers Ltd., New Delhi) fitted with Porapak R column and flame ionization detector. Nitrogen was used as the carrier gas. Nitrogenase activity was expressed in μmol C_2H_4 formed mg protein^{-1} h^{-1}.

Indole-3-acetic acid (IAA) production : The production of indole acetic acid (IAA) was tested as per the method of Gordon and Weber (1951). Briefly, 1.5 ml bacterial culture grown in medium supplemented with tryptophan (100 μg/ml) for 72 h was centrifuged at 12,000 rpm for 5 min and 1 ml supernatant was retained. To this, 2 ml $FeCl_3$-$HClO_4$ reagent was added and after 25 min absorbance of pink color appearing, if any, was measured at 530 nm in a Milton Roy 1201 UV-Vis Spectrophotometer (Milton Roy Company, NY, USA).

Test for Phosphate solubilization : Test for phosphate solubilization was made as per the method of Goldstein (1986). The presence of clearing zone around bacterial colonies after 96 h of growth at 30°C was used as indicator for positive P solubilization.

Siderophore assay : Production of siderophore was tested qualitatively following the method of Neilands (1981). Accordingly, CAS agar plates were prepared which contained: 1mM Chromeazurol-S, 10 ml $FeCl_3.6H_2O$ (1mM final concn.) made in 10 mM HCl, and N, N-cetyl trimethyl ammonium bromide (2 mM) (CTAB). Spot inoculation of test culture was made on CAS-agar plates and incubated at 30°C for 72 h. Yellow to orange halo zone appearing around the colonies was recorded as positive test for siderophore production.

Estimation of 1-aminocyclopropane-1-carboxylate deaminase (ACC deaminase) activity : ACC deaminase activity was estimated following the method described by Penrose and Glick (2003).

Amplification of 16S rDNA : Genomic DNA was extracted employing DNEasy extraction kit (Qiagen, Germany) as per the instructions of the manufacturer. 1.5 kb 16S rRNA gene was amplified using universal primer in a final volume of 50 μL The PCR reaction mix included; 1.5 U of *Taq* DNA polymerase (Bangalore Genei, India), 1 X PCR buffer with 1.5 mM $MgCl_2$, 50 pmol of each forward and reverse primers (Integrated DNA Technologies, Inc., IA, USA), 125 μM of each dNTPs (Bangalore Genei, India) and 50 ng template DNA. The sequence of primer used for amplification was 8f 5'-AGA GTT TGA TYM TGG CTC AG-3'and 1495r 5'-CTA CGG CTA CCT TGT TAC GA-3' where Y = C or T and M = A or C. Thermal program for the amplification was set as: initial denaturation at 94°C for 3 min, 35 cycles of 1 min at 94°C, 1 min at 58°C and 2 min at 72°C followed by final extension at 72°C for 5 min and storage at 4°C. Partial 16S rRNA of V2-V3 region (400 bp) was amplified by the primer set; Fd 5'-ACT GGC GGA CGG GTC AGT AA-3' and rev. 5'-CGT ATT ACC GCG GCT GCT GG-3'. Thermal profile was set as: an initial denaturation at 95°C for 3 min followed by 30 cycles of 95°C for 1 min, 50°C for 1 min, 72°C for 1 min. 10 s, and final extension at 72°C for 5 min.

Amplified ribosomal DNA restriction analysis (ARDRA) : 1.5 kb 16S rDNA amplified by PCR was purified and subjected to restriction digestion by *Alu*I and *Rsa*I restriction enzymes as per the instructions of manufacturer (New England Biolabs Ltd, Hertfordshire, England, UK). Restriction digestion was done in a final volume of 25 μl containing 1 X restriction enzyme buffer, 0.125 μl (1.25 U) restriction enzyme and 15 μl PCR product. After mixing the reaction mixture carefully, samples were incubated for 3 h in a water-bath preset at 37°C. Reaction was terminated by heat inactivation of restriction enzymes at 70°C for 20 min .The samples were analyzed by agarose gel (3%) electrophoresis and monitored on gel documentation unit.

Denaturing gradient gel electrophoresis (DGGE) : 16S rDNA (200 bp with GC clamp) was amplified with the forward primer containing GC clamp at 5' end (F352T: 5'- CGC_CCG CCG CGC GCG GCG GGC GGG GCG GGG GCA CGG GGG GAC TCC TAC GGG TGG C- 3' and 519r: 5'-ACC GCG GCT GCT GGC AC- 3') in 25 μl of reaction mixture containing 1×PCR buffer, 2.5mM $MgCl_2$ (Bangalore Genei, India), 100 ng of the template DNA, 25.0 pmol each of the forward and reverse primers, 250 ìM each of dNTPs, and 1 U of *Taq* DNA polymerase (Bangalore Genei, India). The touchdown PCR program was performed which consisted of an initial denaturation at 95°C for

5 min, followed by 6 cycles of 95°C for 1 min, 65°C for 1 min, and 72°C for 1 min, in which the annealing temperature was reduced by 0.5°C/cycle from the preceding cycle, and then 24 cycles of 95°C. Perpendicular DGGE was performed with "The Decode Universal Mutation Detection System" (Bio-Rad Laboratories, USA). A uniform gradient gel of 0% to 100% denaturant was prepared which was changed several times so as to optimize suitable concentration and finally 20 to 60% denaturant was found optimal for the best result.

16S rDNA sequencing : Sequencing of 16S rDNA was done with forward primer from both the ends with ABI Prism (Model 3100) automated DNA sequencer using Big Dye- Terminator kit v.3. The sequences obtained were compared with nucleotide sequences present in GenBank using the BLASTn program (Altsch.ul *et al* 1997). Sequence of desired genes (16S rDNA) was analyzed using BLASTn program of NCBI (http://blast.ncbi.nlm.nih.gov/Blast.cgi). Neighbor-joining phylogenetic tree with kimura method (1980) was constructed in ClustalW (version 1.83) and bootstrapping for 1000 interactions was done.

Results

Initially, standard enrichment technique was used to enrich diazotrophic bacteria present in the rice rhizosphere of five districts of eastern UP (Table 1). Accordingly, JNFb$^-$ medium (devoid of inorganic combined nitrogen sources) was used which resulted in the isolation of 295 isolates from 40 rhizospheric soil samples of five districts. Since several isolates were found common in all the districts, such isolates were excluded and 99 isolates having distinct morphotypes in terms of colony morphology, size, colour, elevation and mucilage formation were finally selected (Table 1). Standard serial dilution technique was employed for MPN which revealed population density in the range of 3.34x 10^6- 1.33x10^7CFU/gm of soil in samples of all the five districts. Test of growth in nitrogen-free semi-solid medium (0.15% agar-agar) showed the formation of pellicles by majority of the isolates. However the colour of pellicles varied from whitish to creamy to yellowish and marked differences in the location of pellicles were observed (1–2 mm below the surface of the medium in some isolates and 4–12 mm in others) in different isolates.

Preliminary screening of plant growth promoting (PGP) characters namely, IAA production, P solubilization and siderophore production showed the presence of one or other PGP characters in all the 99 isolates.

Table 1 : Distribution pattern of diazotrophic bacteria in rice rhizosphere of eastern UP

Districts	Types of total isolates	Distinct isolates	IAA - producers	P-solubilizers	Siderophore-producers
Chandauli	38	11	9 (81.81%)	2 (18.18%)	3 (27.27%)
Ghazipur	71	23	16 (69.56%)	5(21.73%)	10 (43.47%)
Jaunpur	69	25	22 (88.00%)	5 (20.00%)	6 (24.00%)
Mirzapur	23	7	3 (42.85%)	2 (28.57%)	2 (28.57%)
Varanasi	94	33	21 (63.63%)	8 (24.24%)	16 (48.48%)
Total	295	99	71(71.71%)	22(22.22%)	37(37.37%)

Assay of nitrogenase activity showed different levels of acetylene reduction activity among all the isolates. Out of 99 isolates, the highest rate of nitrogenase activity was noted in the isolate V1S7 (1.72 μmol C_2H_4 mg protein^{-1} h^{-1}) followed by G1S3 (1.65 μmol C_2H_4 mg protein^{-1} h^{-1}). In general, the rate of nitrogenase activity ranged between 1.72 to 1.33 μmol C_2H_4 mg protein^{-1} h^{-1}. Screening for IAA production revealed positive test for 71 (71.71%) isolates. Among 71 isolates, the highest production of IAA was noted in the isolate J2S5 (353 μg IAA mg protein^{-1}) and the lowest level was noted in G4S10 (10.10 μg IAA mg protein^{-1}). Details of IAA- producing isolates are given in Table 1. Interestingly, none of the isolates showed IAA production in the medium lacking tryptophan (100μg/ml). Test of P solubilization revealed positive result for 22 (22.22%) isolates out of 99 isolates. Similarly of the 99 isolates, 37 (37.37%) isolates were found to produce siderophore on CAS plate (Table 1). Assay of ACC deaminase revealed that the level of activity ranged from 28.15 to 294 nmol a-ketobutyrate mg protein^{-1}h^{-1} in various isolates. It was interesting to note that all the isolates showing ACC deaminase activity had also very high level of nitrogenase activity (1.33 to 1.72 mmol C_2H_4 mg protein^{-1} h^{-1}) suggesting their potential role as biofertilizer.

Once it became evident that all the 99 isolates show significant differences in morphological characters, growth pattern and PGP characters, it was desirable to analyze molecular diversity employing PCR assay. To achieve this objective 16S rDNA (1.5 kb) was successfully amplified from all the 99 isolates and ARDRA was performed. Restriction patterns generated by the use of *Alu*1 and *Rsa*1 showed marked differences in all the 99 isolates. Typical representation of ARDRA profile of all the isolates (23) of Ghazipur district is shown in Fig.1A & B. It is also evident from the restriction pattern generated by two enzymes that the *Alu*I shows better distinguishing pattern (Fig.1A) than *Rsa*1 (Fig.1B). With a view to reveal homology between all the isolates of Ghazipur, combined *Alu*1 and *Rsa*1 restriction pattern of ARDRA was used for cluster analysis using NTSYS PC 2.1 software (Fig. 2). It is evident from the results that all these isolates are grouped in three clusters. Clusters A, B and C consist of 2, 19 and 2 isolates respectively and none of the isolates showed 100% homology. Cluster A showed 68% similarity with cluster B and C whereas cluster B and C showed 70.60% similarity to each other (Fig. 2). That the isolates recovered from all the five districts are indeed different was further confirmed by DGGE. It is evident form the DGGE profile of all the isolates of Varanasi district that there are significant differences in position of desired band among all the isolates (Fig.3).

Among all the isolates tested, isolates V1S7, J1S9, M2S7, J2S10, G1S3, M2S10 and C3S5 showed highest level of nitrogenase activity, IAA production and siderohpore production and P solubilisation, this aroused our interest to characterize and identify these potential isolates. Accordingly, full length 16S rDNA (1.5 kb) of C3S5 and partial 16SrDNA (400bp) of V1S7, J1S9, M2S7, J2S10, G1S3 and M2S10 were amplified and sequenced. On the basis of sequence homology available in GenBank, these isolates were identified as *Sphingomonas sp.* C3S5 (FJ012066), *Pantoea agglomerans* V1S7 (EU040249), *Bacillus megaterium* J1S9 (EU040252), *Bacillus megaterium* M2S7 (EU040253), *Microbacterium sp.* J2S10 (EU040255), *Bacillus megaterium* G1S3 (EU040256), and *Cronobacter turicensis* M2S10 (EU040257) respectively.

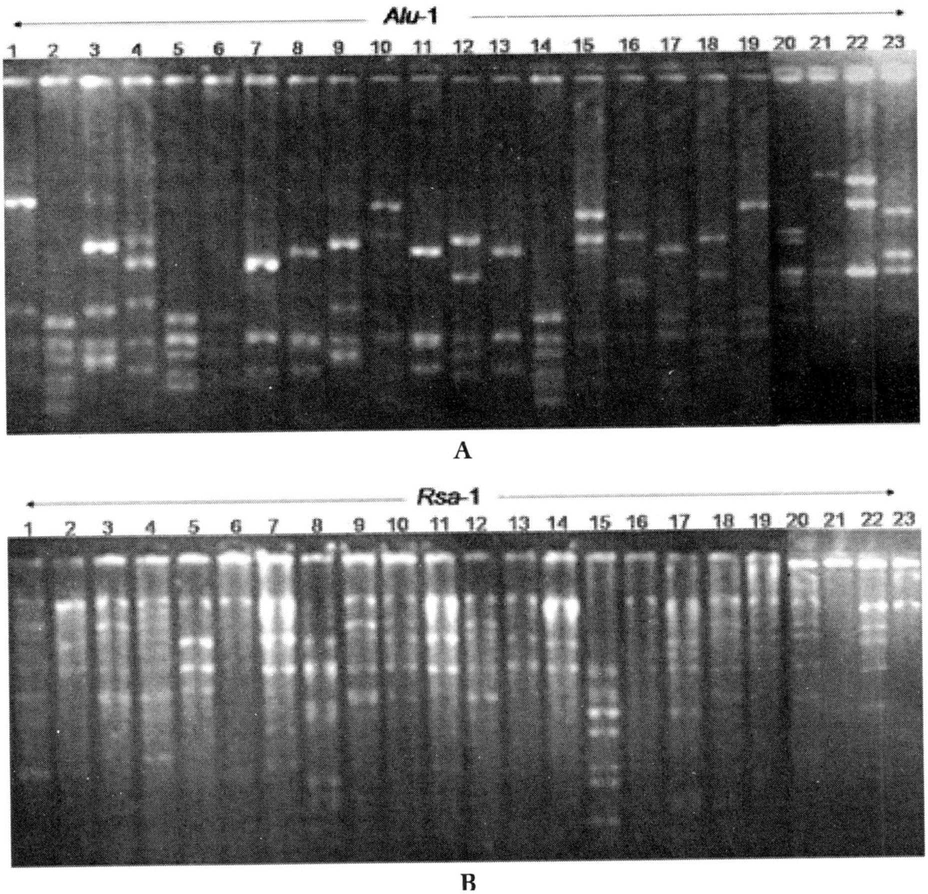

Figure 1 : ARDRA profile of various isolates of Ghazipur district. 16S rDNA (1.5 kb) was digested with *Alu1* (A) and *Rsa1*(B). Lanes 1 to 23; 1- G1S3, 2- G3S10, 3- G3S8A 4- G3S12, 5- G4S10, 6- G6S10, 7- GS1S1, 8- GS1S2, 9- GS1S9, 10- GS3S3, 11- GS4S4, 12-GS7S1, 13-GS8S2, 14-GS8S5, 15-G4S11, 16-GSE, 17-G5S9, 18- GS2S6, 19- GS7S4, 20-G6S6, 21-GS3S1, 22-GS3S5, and 23-GS4S2. G-Ghazipur-S-site number and, S-Bacterial sample number.

Knowing the taxonomic assignment of selected isolates, it was desirable to find out homology status of these isolates with other species available in data base. Accordingly phylogenetic tree was constructed using MEGA4 after aligning the sequences with Clustal W (Fig.4). The evolutionary history was inferred using the Neighbor-Joining method. Isolates reported in this study showed close homology with respective species/genus of the data base.

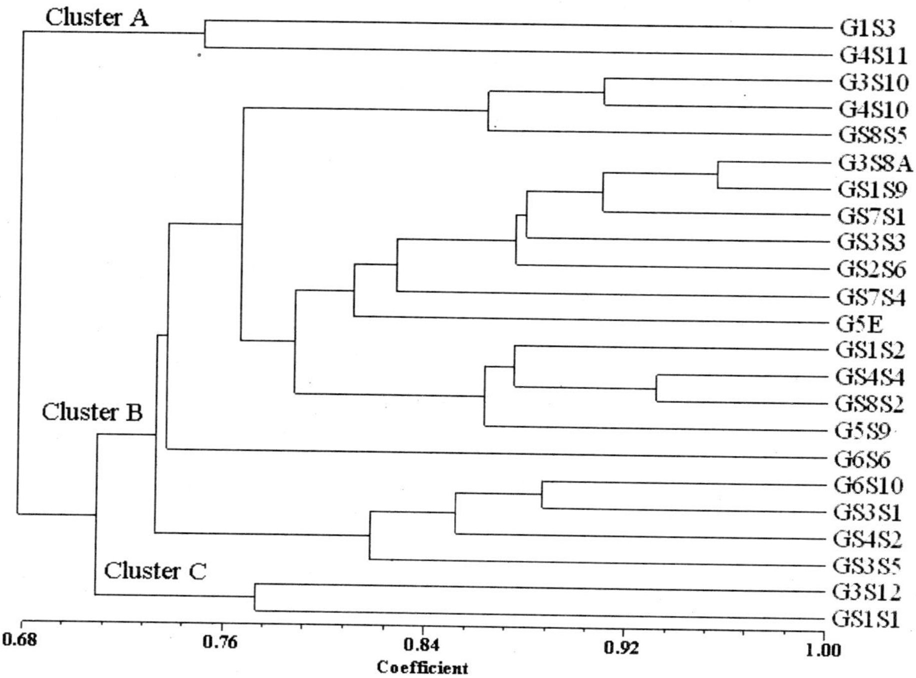

Figure 2 : Dendrogram based on restriction digestion pattern of 16S rDNA (1.5 kb) digested with *Alu1* and *Rsa1*. Phylogenetic tree was constructed using NTSYS PC 2.1 software.

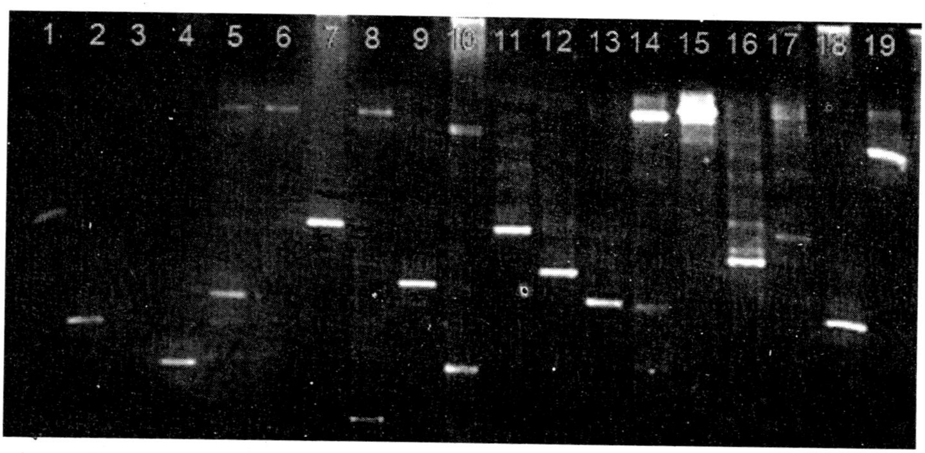

Figure 3 : DGGE of 16S rDNA amplified product (200 bp) from 19 isolates of Varanasi district: Lanes 1-19: 1-VB4S7A, 2- VB7S4, 3-VA7S7, 4-VB5S10, 5-VB4S6, 6-VB1S2, 7- V1S6, 8-VB3S1, 9-VA8S1, 10-VA1S6A, 11-VB5S5, 12-VB7S5, 13-VA2S3, 14- VA7S1, 15- VA3S1, 16-VA1S2, 17- VB1S5, 18- VB5S7, and 19-VA4S7. V-Varanasi, B-Babatpur site, and S- Bacterial sample number.

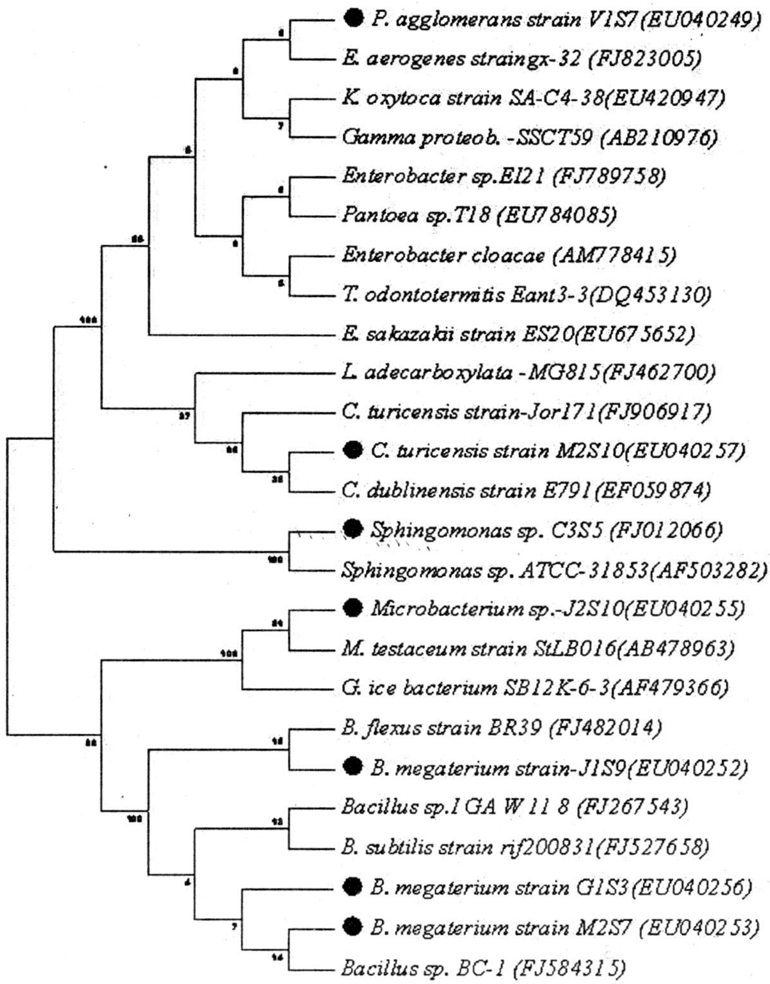

Figure 4 : Phylogenetic analysis of seven identified bacterial isolates.The evolutionary history was inferred using the Neighbor-Joining method. The bootstrap consensus tree inferred from 1000 replicates is taken to represent the evolutionary history of the taxa analyzed. The evolutionary distances were computed using the Kimura 2-parameter method and are in the units of the number of base substitutions per site.

Discussion

The plant rhizosphere is a versatile and dynamic ecological environment of intense microbes- plant interactions for harnessing essential micro- and macro- nutrients from a limited nutrient pool

(Jeffries *et al.* 2003). The above notion is reflected from the fact that we have successfully isolated 295 isolates from rice rhizospheres of five districts of eastern UP. However it is evident from our results that several isolates are common in all the rice fields since detailed characterization could confirm the presence of only 99 distinct morphotypes. Differences in types of bacterial isolates in rice rhizosphere of individual districts are noteworthy and this may be due to the differences in soil types and/or agro-climatic conditions. Our results are in agreement with previous reports where vast differences in bacterial assemblages were noted in a particular plant species growing at different locations (De Weger *et al.* 1995; Ladha and Reddy 2000; Roesch *et al.* 2007).

Enrichment technique applied in this study proved valid since all the isolates exhibiting diazotrophic growth also showed appreciable level of nitrogenase activity. This clearly suggests that these isolates convert atmospheric dinitrogen (N_2) to ammonia which in turn serves as the sole source of nitrogen for growth and development. Rhizospheres of several plants have been reported to be the preferred niche for the growth of diazotrophic bacteria (Ladha and Reddy 2000). As such the preference of bacteria for rhizospheric association is thought to be mainly due to low tension of oxygen in root zone. Additionally, metabolites secreted by the roots become readily available to the associated bacteria (Franche *et al.* 2009). Exudates of roots contain a variety of organic carbon/ N compounds which favor growth of certain bacteria. Most probably this might also be one of the reasons for the presence of different types of bacteria in rhizosphere of a particular plant species.

Multiple PGP characters among rhizobacteria have been reported by other workers (Ahmad *et al.* 2008) while such reports in indigenous isolates of India particularly from Eastern UP are less common. Interestingly our results clearly demonstrated the presence of one or more than two PGP characters in all the 99 isolates. Similar to the findings of other workers (De Weger *et al.* 1995; Ahmad *et al* .2008), the frequency of IAA-producers was highest followed by siderophore producers and P solubilizer. Out of 99 isolates, 71 were capable to produce IAA in the range of 10 to 353 μg IAA mg protein^{-1} in the presence of tryptophan. In addition to the differences observed in the level of IAA production among different isolates, the percentage of IAA- producing strains was also significantly different in different sites. Pattern of P solubilization and siderophore production in terms of rate and distribution among different isolates was more or less similar to that of IAA producers. Whereas the differences in the rate of IAA production or other PGP characters are surely governed by the genetic set up of an individual isolate, the percentage of PGP isolates existing in any site seems to be due to the nature of soil types and/or nutrient status. Soil rich in organic matter especially due to litter decomposition shall surely favor the growth of IAA-producers since the chances of tryptophan availability are expected to increase.

Reports showing production of metabolites beneficial to plants, such as, phytohormones, siderophores, certain antibiotics and P solubilization by certain isolates have created confusion about the specific role of a particular PGP trait (Kennedy *et al.* 2004; Roesch *et al.* 2007). However, there are a number of reports which have clearly demonstrated that bacteria having any one of the PGP characters show plant growth enhancement (Ladha and Reddy 2000; Vessey 2003). For example, several isolate improve the supply of P to plants as a consequence of their ability to solubilize insoluble P, certain isolates cause root elongation by IAA production and a few which produces siderophores are responsible for increased availability of iron. Henceforth a single PGPR does not need to possess all PGP characteristics and plant growth promotion can result from a single mechanism.

ACC-deaminase activity of PGPR strains is known to enhance the root length and growth by sequestering and hydrolyzing ACC from germinating seeds and there by increasing the activity of rhizosphere zone (Glick *et al.* 1998). An ACC-deaminase lacking mutant of *E. cloacae* UW4 was found not to promote root growth of canola (Li *et al.* 2000). Results of this study clearly showed that various isolates are capable to exhibit ACC deaminase activity and hence may have potential role in plant growth promotion. In this study, altogether seven isolates namely C3S5, G1S3, J1S9, J2S10, M2S7, M2S10 and V1S7 were found to have multiple plant growth-promoting traits. It is worth to examine whether all the traits of PGPR are expressed simultaneously or at different phases of growth. However it is sure that strains endowed with multiple PGPR traits have better chances of being a successful inoculant strain.

DNA-based techniques are well documented to provide a comprehensive measure of the diversity and composition of rhizospheric bacteria (Schmalenberger *et al.* 2001; Miyamoto *et al.* 2004). In general, PCR-based techniques such as RAPD, ERIC-PCR, ARDRA, BOX-PCR and DGGE have been frequently used to study molecular diversity among bacteria from different habitats. We have used ARDRA to investigate diversity among 99 isolates of rhizospheric soil of five districts of eastern UP. ARDRA profile of all the 99 isolates showed distinct banding pattern after restriction digestion with *Alu*1 and *Rsa*1. Additionally, DGGE profile, where a single reproducible band with higher intensity was considered, also showed diversity among different isolates. That these isolates indeed differ among each other became evident from the dendrogram (Fig.2) which showed grouping of all the isolates under different clusters. Based on the data of ARDRA and DGGE profile, it may be concluded that there does exist significant molecular diversity among 99 isolates, a fact which is also supported by morphological characteristics and PGP characters of the isolates reported herein.

Since our aim was to screen and identify the most efficient PGPR, we successfully identified 7 isolates namely C3S5, G1S3, J1S9, J2S10, M2S7, M2S10 and V1S7 as *Sphingomonas sp.*C3S5, *Bacillus megaterium* G1S3, *Bacillus megaterium* J1S9, *Microbacterium sp.* J2S10, *Bacillus megaterium* M2S7, *Cronobacter turicensis* M2S10 and *Pantoea agglomerans* V1S7 respectively. The identification was based on 99%~100% similarity of these sequences to those reported in GenBank. To our knowledge, this is the first comprehensive study on PGPR which was made on rice rhizosphere from eastern Uttar Pradesh and the results obtained may be exploited for the development of efficient biofertilizer strains.

Acknowledgements

This work was supported by a research grant sanctioned to AK by ICAR (No. NBAIM/AMAAS/MD (19)/AK/BG/3), New Delhi. Ashutosh Kumar Rai and Shailesh Kumar Shahi are working on the post of SRF in this project.

References

Ahmad F, Ahmad I and Khan MS. (2008). Screening of free-living rhizospheric bacteria for their multiple plant growth promoting activities. *Microbiol Res* 163:173-181.

Altschul SF, Madden TL, Schafeer AA, Zhang J, Zhang Z, Miller W and Lipman DJ. (1997). Gapped BLAST and PSI-BLAST: A new generation of protein database search programs. *Nucleic Acid Res* 25:3389-3402.

Bottini R, Cassan F and Piccoli P. (2004). Gibberellin production by bacteria and its involvement in plant growth promotion and yield increase. *Appl. Microbiol. Biotechnol.* 65: 497-503.

De Freitas JR, Banerjee MR and Germida JJ. (1997). Phosphate-solubilizing rhizobacteria enhance the growth and yield but not phosphorus uptake of canola (*Brassica napus L.*). *Biol Fertil Soils* 24: 358-364.

De Weger LA, van der Bij AJ, Dekkers LC, Simons M, Wijffelman CA and Lugtenberg BJJ. (1995). Colonization of the rhizosphere of crop plants by plant-beneficial Pseudomonads. *FEMS Microbiol Ecol* 17: 221-228.

De Los Santos PE, Bustillos-Cristales R and Caballero-Mellado J. (2001). *Burkholderia,* a genus rich in plant-associated nitrogen fixers with wide environmental and geographic distribution. *Appl Environ Microbiol* 67: 790- 798.

Franche C, Lindstrom K and Elmerich C. (2009). Nitrogen-fixing bacteria associated with leguminous and non-leguminous plants. *Plant Soil* 321: 35-59.

Glick BR, Penrose DM and Li J. (1998). A model for the lowering of plant ethylene concentrations by plant growth promoting bacteria. *J Theor Biol* 190: 63–68.

Glick BR. (1995). The enhancement of plant growth by free-living bacteria. *Can J Microbiol* 41: 109–114.

Goldstein AH. (1986). Bacterial solubilisation of mineral phosphates: historical perspective and future prospects. *Am J Alter Agric* 1: 51-57.

Gordon SA and Weber RP. (1951). Colorimetric estimation of indole-acetic acid. *Plant Physiol* 26:192-195.

Jeffries P, Gianinazzi S, Perotto S, Turnau K and Barea JM. (2003). The contribution of arbuscular mycorrhizal fungi in sustainable maintenance of plant health and soil fertility. *Biol Fertil Soils* 37: 1–16.

Kennedy IR, Choudhury ATMA and Kecskes ML. (2004). Non-symbiotic bacterial diazotrophs in crop-farming system: can their potential for plant growth promotion be better exploited? *Soil Biol Biochem* 36: 1229-1244.

Kimura M. (1980). A simple method for estimating evolutionary rate of base substitutions through comparative studies of nucleotide sequences. *J Mol Evol* 16: 111-120.

Kloepper JW, Lifshitz R and Zablotowicz RM. (1989). Free-living bacterial inocula for enhancing crop productivity. *Trends Biotechnol* 7: 39–43.

Ladha JK and Reddy PM. (2000). The quest for nitrogen fixation in rice. Proceedings of the Third Working Group Meeting on Assessing Opportunities for Nitrogen Fixation in Rice, 9–12 Aug. 1999, IRRI, Los Banos, Laguna, Philippines, 354pp.

Li J, Daniel H, Charles TC and Glick BR. (2000). An ACC deaminase minus mutant of Enterobacter cloacae UW4 no longer promotes root elongation. *Curr Microbiol* 41: 101–105.

Miyamoto T, Kawahara M and Minamisawa K. (2004) . Novel endophytic nitrogen-fixing Clostridia from the grass *Miscanthus sinensis* as revealed by terminal restriction fragment length polymorphism analysis. *Appl Environ Microbiol* 70: 6580-6586.

Neilands JB. (1981). Microbial iron compounds. *Annu Rev Biochem* 5: 715-731

Pedraza RO, Ramirez-Marta A, Xiqui ML and Baca BE. (2004). Aromatic amino acid aminotransferase activity and indole-3-acetic acid production by associative nitrogen-fixing bacteria. *FEMS Microbiol Lett* 233: 15-21.

Penrose DM and Glick BR. (2003). Methods for isolating and characterizing ACC deaminase-containing plant growth promoting rhizobacteria. *Physiologia Plantarum* 118: 10–15.

Roesch LFW, Camargo PDFAO and Triplett EW. (2007). Screening of diazotrophic bacteria *Azopirillum* spp. for nitrogen fixation and auxin production in multiple field sites in southern Brazil. *World J Microbiol Biotechnol* 23: 1377-138.

Schmalenberger A, Schwieger F and Tebbe CC. (2001) . Effect of primers hybridizing to different evolutionay conserved regions of the small-subunit rRNA gene in PCR-based microbial community and genetic profiling. *Appl Environ Microbiol* 67: 3357-3363.

Stewart WDP, Fitzgterald GP and Burris RH. (1967). *In situ* studies on N_2 fixation using acetylene reduction techniques. *Proc Natl Acad Sci* USA 58: 2071–2078.

Van der Lelie D, Barac T, Taghavi S and Vangronsveld J. (2005). Response to new man: New uses of endophytic bacteria to improve phytoremediation. *Trends Biotechnol* 23: 8-9.

Vessey JK. (2003). Plant growth promoting rhizobacteria as biofertilizers. *Plant Soil* 255: 571–586.

□□□

Chapter 7

APPLICATION OF BIOFORMULATIONS OF PGPRs FOR PLANT GROWTH PROMOTION IN TEA SEEDLINGS

A.P.CHAKRABORTY, B.N.CHAKRABORTY AND U.CHAKRABORTY

Immuno-Phytopathology Laboratory, Department of Botany, University of North Bengal, Siliguri- 734 013, West Bengal, India

Introduction

Use of microorganisms for plant growth promotion and disease control is well recognized (Glick 1995; whipps 2001). Importance of native strains and ecological specificity while selecting the microbial inoculants for a specific environment is also realized (Pandey *et al.* 1998). Isolation of microorganisms, screening for desirable characters, selection of efficient strains and production of inoculums are important steps for making use of this microbe-based technology. One of the common means of application of bacterial inoculants to soil is in the form of bioformulations. Use of plant growth promoting microbes as biocontrol agents and biofertilizers give eco-friendly and inexpensive alternatives to the use of chemicals. However, for easy handling of such bacteria, it is necessary to pack such bacteria in inert materials which can also be packaged and stored. Initially, it is essential to determine whether the bacteria can survive in the bioformulations for a reasonable period of time and whether they can induce similar effects to those observed by live bacterial cells. Viability of inoculums in an appropriate formulation for a certain length of time is important for commercialization of the technology (Bashan 1998). The efficacy of bacterial inoculants would largely depend on the type of formulation and delivery technology (Lumsden *et al.* 1995). Experimental formulations of *Bacillus spp* that have effectively reduced plant diseases have included peat and chitin (Manjula and Podile, 2001; Sid Ahmed *et al.* 2003); ca- alginate, alginate manucol (Schmidt *et al.* 2001).

In the present investigation, the two selected bacterial inoculants were prepared in three formulations and tested for their growth promotion through bioassays using tea seedlings as a test species. The bioassays were conducted using fresh as well as stored preparations of the bacterial inoculants. The viability of the bacterial inoculants in all formulations under storage at room temperature was evaluated over nine months. The study was done keeping in mind, its implications in field applications and commercialization of the bacterial inoculants.

Materials and Methods

Bacterial inoculants: *Bacillus megaterium* TRS-4 and *Serratia marcescens* TRS-1 were isolated from the rhizosphere soil of tea bushes from Hansqua Tea Estate, West Bengal, India and preliminarily identified on the basis of morphological, microscopic and biochemical tests. The identity of the bacteria were confirmed from the Plant Diagnostic and Identification Services, UK. The bacteria were maintained on NA slants. To prepare the inoculum, initially *S. marcescens* and *B. megaterium* were cultured in Nutrient Broth medium (Himedia, M002-100G, ingredients- peptic digest of animal tissue- 5.00g/litre, sodium chloride- 5.00g/l, beef extract- 1.50 g/l, yeast extract- 1.50 g/l, final pH 25°C -7.4±0.2) and were allowed to grow properly, with shaking at 37°C at 120 r.p.m for 48 h. At the end of the log phase, bacterial cultures were centrifuged at 10,000 r.p.m for 15 mins and the supernatant was discarded, selecting the bacterial pellet. Pellet was scraped into sterile distilled water. The aqueous suspensions were diluted as necessary to maintain the bacterial concentration at 10^{10} c.f.u/ml.

Formulations : The bacterial inoculants were prepared in following three formulations: (1) saw dust, (2) rice husk and (3) tea waste.

For the preparation of bioformulations, 2.5 g of carboxy methyl cellulose sodium salt (Himedia) was added to 250 g of saw dust or rice husk and pH was adjusted to 7 by adding calcium carbonate. They were sterilized for 30 min in two consecutive days. 100 ml of aqueous bacterial suspensions (a concentration of 1×10^{10}c.f.u/ml) was added to the mixtures under sterile condition. The mixture was dried under shade to reduce the moisture to less than 20%. Formulations were packed in polythene bags, sealed and stored at room temperature. In preparation of tea waste bioformulation, the procedure was similar but tea waste was initially soaked overnight in distilled water to remove the phenolic components. All the formulations were used as fresh and after 9 months storage at room temperature under laboratory conditions.

Bioassay for evaluation of growth promotion and rhizosphere colonization by bacterial inoculants in various formulations : The bioassay was conducted using different varieties of tea (TV-18, TV-23, TV-25, TV-26 and T-17) as a test species. The selected tea seedlings were maintained in 12½ earthenware pots in randomized block design. Tea seedlings were watered regularly for proper maintenance. 100 ml of aqueous bacterial suspensions at 10^9-10^{10} cfu/ml were added to sterilized 250g of saw dust, rice husk or tea waste along with 2.5 g of carboxy methyl cellulose as a carrier and calcium carbonate under sterile condition. To test *in vivo* efficacy of the three bioformulations for plant growth promotion, 2-mo-old as well as fresh formulations were applied in potted soil at the rate of 100g/pot.

For determining growth promotion of tea seedlings followed by treatments, increase in height, number of leaves were noted in potted plants. Plants were grown under natural conditions of light and temperature (30±2°C). Observations were recorded after 2 months of application. In each plot, for each treatment 10 replicates were taken and average of the 10 replicate plants were analysed.

Enumeration of viable bacteria in formulations under storage : Viable bacteria in the formulations used in the study were conducted by dilution-plate technique using Nutrient Agar medium. The plates were inoculated at 37°C and cfu were conducted. The viability tests were carried out with the fresh as well as stored preparations. Viability was checked at an interval of one months till nine months. The data was expressed in log form for analysis.

Results

Effect of S.marcescens and B. megaterium in bioformulatrions on plant growth : Application of *S.marcescens* and *B. megaterium* in bioformulations of saw dust, rice husk and tea waste led to significant increase in growth of tea seedlings (**Tables 1, 2 &3**). Better

Table 1: *In vivo* application of bioformulation of *S. marcescens* in potted plants

Tea varieties	Treatment	2 months after treatment			
		Initial height	Final Height	Initial leaf numbers	Final leaf numbers
TV-18	Control	19±1.3	26± 1.5 (7)	19.3±1.8	24.5± 1.2(5.2)
	Serratia marcescens	22±2.1	30.2 ± 1.8(8.2)	19.1±1.6	25.5± 1.4(6.4)
	S.m + saw dust	21.2±2.0	28.3 ± 1.7(7.1)	18.5±0.9	26± 1.4(7.5)
	S.m + rice husk	20±1.5	31± 2.1(11)	20.4±1.1	26.1 ±0.9 (5.7)
	S.m + tea waste	19±1.0	27.2 ± 1.9(8.2)	21.3±0.8	27.5± 1.1(6.2)
TV-23	Control	22.4±0.9	25± 1.7(2.6)	17.1±1.2	24± 1.0(6.9)
	Serratia marcescens	24.5±1.6	32.2 ± 2.1(7.7)	18.4±1.3	25.2±1.7 (6.8)
	S.m + saw dust	24.2±1.4	37.3 ± 2.4(13.1)	18.5±1.4	25± 1.3(6.5)
	S.m + rice husk	23.1±1.6	35± 1.8(11.9)	19.6±0.8	26.2± 2.0(6.6)
	S.m + tea waste	22±1.9	35.5 ±2.0 (13.5)	18.3±1.6	26±1.5 (7.7)
TV-25	Control	21.4±2.0	24.5 ±1.9 (3.1)	19.5±1.0	25.2±1.8 (5.7)
	Serratia marcescens	22.5±2.3	27.5 ± 2.0(5.0)	20±1.8	28± 1.1(8.0)
	S.m + saw dust	23.3±1.9	33± 1.8(9.7)	20.2±1.9	29.1± 2.0(8.9)
	S.m + rice husk	22.2±1.7	35.3 ± 1.7(13.1)	21.4±1.2	27.2± 1.6(5.8)
	S.m + tea waste	21±0.9	29± 1.4(8.0)	22.3±2.0	28.5± 1.1(6.2)
TV-26	Control	22.4±1.8	25.5 ±2.3 (3.1)	17.5±2.0	21± 1.5(3.5)
	Serratia marcescens	23.2±1.4	28.5 ± 2.1(5.3)	18.2±1.9	30.3± 1.3(12.1)
	S.m + saw dust	24.5±1.5	28.2 ± 2.4(3.7)	19.4±1.5	31.1± 1.0(11.7)
	S.m + rice husk	23±1.2	27.2 ± 2.7(4.2)	18.1±1.7	29± 1.9(10.9)
	S.m + tea waste	23.2±1.8	27.5 ± 1.8(4.3)	19.2±1.5	31± 1.7(11.8)
T-17	Control	24±2.4	27.2 ± 1.2(3.2)	18.5±1.5	20.4±1.8 (1.9)
	Serratia marcescens	24.2±2.2	33.1 ± 1.7(8.9)	18.2±1.9	25.5± 1.0(7.3)
	S.m + saw dust	23.5±2.1	33.2 ± 2.0(9.7)	19±1.2	26± 1.1(7.0)
	S.m + rice husk	25.2±2.0	34.2 ± 1.4(9.0)	19.2±1.7	27.2± 1.4(8.0)
	S.m + tea waste	26.1±2.3	32.5 ± 1.1(6.4)	19.5±1.8	25.3± 1.1(5.8)
CD (P=0.05)	(Treatments)	1.233	2.875	0.892	2.186
	(Varieties)	1.233	2.875	0.892	2.186

Table 2: *In vivo* application of bioformulation of *B. megaterium* in potted plants

Tea varieties	Treatment	2 months after treatment			
		Initial height	Final Height	Initial leaf numbers	Final leaf numbers
TV-18	Control	19±1.3	26± 1.5 (7)	19.3±1.8	24.5± 1.2(5.2)
	Bacillus megaterium	25±1.5	30±2.1 (5)	19±1.3	26±2.0(7.0)
	B.m + saw dust	20±1.9	27± 1.2(7)	17±0.9	24± 1.8(7.0)
	B.m + rice husk	19.5±1.5	32± 1.1(12.5)	18.1±1.0	28± 2.1(9.9)
	B.m + tea waste	21±1.8	29±1.8 (8.0)	22±1.2	27±2.4(5.0)
TV-23	Control	22.4±0.9	25± 1.7(2.6)	17.1±1.2	24± 1.0(6.9)
	Bacillus megaterium	24.5±1.5	33±2.1(8.5)	19±1.1	28±1.9 (9.0)
	B.m + saw dust	24±1.4	38± 2.5(14)	21±1.7	32± 2.3(11)
	B.m + rice husk	23±2.1	39± 2.7(16)	18±1.0	31± 2.4(13)
	B.m + tea waste	22±2.2	36± 1.9(14)	19.5±2.0	30.2±1.4(10.7)
TV-25	Control	21.4±2.0	24.5±1.9 (3.1)	19.5±1.0	25.2±1.8 (5.7)
	Bacillus megaterium	29±1.9	32±2.1 (3.0)	34.2±2.0	45.0±1.1(10.8)
	B.m + saw dust	27±2.1	29± 1.3(2.0)	34±2.1	45.2± 2.1(11.2)
	B.m + rice husk	26.1±2.0	28.1± 1.1(2.0)	31.5±1.9	42.5±1.8 (11.0)
	B.m + tea waste	28±2.2	32± 2.1(4.0)	30.2±1.3	41±11.9(10.8)
TV-26	Control	22.4±1.8	25.5±2.3 (3.1)	17.5±2.0	21± 1.5(3.5)
	Bacillus megaterium	25±1.7	33± 1.7(8.0)	19±1.5	32±2.2 (13)
	B.m + saw dust	22±1.1	29.2± 1.1(7.2)	17±1.0	30±2.4(13)
	B.m + rice husk	21±1.7	31± 1.8(10)	16±1.2	27± 1.9(11)
	B.m + tea waste	21.2± 1.9	30.2± 1.2(9.0)	18±1.1	26.2± 1.7(8.2)
T-17	Control	24±2.4	27.2± 1.2(3.2)	18.5±1.5	20.4±1.8 (1.9)
	Bacillus megaterium	25±1.5	34± 2.1(9.0)	18±1.6	26±1.5(8.0)
	B.m + saw dust	22±1.4	33± 2.3(11)	21±2.0	26.5±1.7(5.5)
	B.m + rice husk	25.2±1.3	32.5±2.2 (7.3)	19±1.6	27±1.8(8.0)
	B.m + tea waste	27±2.0	31±1.9 (4.0)	20±1.5	28.3±1.1 (8.3)
CD (P=0.05)	(Treatments)	2.58	2.98	3.84	2.85
	(Varieties)	2.58	2.98	3.84	2.85

growth responses were observed in TV-23, T-17 in comparison to other varieties. The growth was measured in terms of increment in height and number of leaves of seedlings two months after application. The observed increase was similar to that observed by application of aqueous suspension of the bacterium. Statistical analysis (ANOVA) revealed that there was no significant difference among the aqueous suspension or different bioformulations though all of them were significantly higher than control.

Table 3: *In vivo* application of combined bioformulation of *B. megaterium* and *S. marcescens* in potted plants

Tea varieties	Treatment	2 months after treatment			
		Initial height	Final Height	Initial leaf numbers	Final leaf numbers
TV-18	Control	19±1.3	26± 1.5 (7)	19.3±1.8	24.5± 1.2(5.2)
	B.megaterium+ S.marcescens	28±1.4	33± 2.1(5.0)	21±1.6	27.2±1.1(6.2)
	B.m+ S.m + saw dust	23±1.4	26.2± 1.6(3.2)	20±1.1	23.5±1.4(3.5)
	B.m + *S.m* + rice husk	19±1.8	30±1.5(11)	18±1.7	26.6±1.9(8.6)
	B.m + *S.m* + tea waste	22±2.0	27±1.3(5.0)	25±2.1	28±1.8(3.0)
TV-23	Control	22.4±0.9	25± 1.7(2.6)	17.1±1.2	24± 1.0(6.9)
	B.megaterium + *S.marcesc ens*	25±1.9	31±2.1(6.0)	21±1.7	29.9±1.1(8.9)
	B.m+ S.m + saw dust	22±1.8	36.7±2.0(14.7)	20±1.6	31±1.2(11)
	B.m + *S.m* + rice husk	23±2.0	37±1.9(14)	18.5±2.1	27.7±2.0(9.2)
	B.m + *S.m* + tea waste	22.1±1.5	35.5±2.3(13.4)	19±1.9	30±1.5(11)
TV-25	Control	21.4±2.0	24.5±1.9 (3.1)	19.5±1.0	25.2±1.8 (5.7)
	B.megaterium + *S.marcescens*	22.5±2.0	29.8±2.1(7.3)	21±2.0	35.5± 2.1(14.5)
	B.m+ S.m + saw dust	21 .2±1.7	28.3±1.9(7.1)	23.2±1.8	35.2±1.8(12.01)
	B.m + *S.m* + rice husk	22±1.0	31±1.7(9.0)	21±1.7	33.4±2.2(12.4)
	B.m + *S.m* + tea waste	25.2±1.8	32.2±1.4(7.0)	23.8±2.0	31.3 1.9(±7.5)
TV-26	Control	22.4±1.8	25.5±2.3 (3.1)	17.5±2.0	21± 1.5(3.5)
	B.megaterium + *S.marcescens*	25.9±1.6	35±2.1(9.1)	19±1.5	37±2.1(18)
	B.m + *S.m* + saw dust	22. 3±2.0	29± 1.8(6.7)	17.7±1.0	34±2.4(16.3)
	B.m + *S.m* + rice husk	18.8±1.9	34±2.1(15.2)	16.3±1.3	26.3±2.5(10.0)
	B.m + *S.m* + tea waste	21±1.1	32.2±1.7(11.2)	15.0±1.1	26±1.8(11)
T-17	Control	24±2.4	27.2± 1.2(3.2)	18.5±1.5	20.4±1.8 (1.9)
	B.megaterium + *S.marcescens*	24±1.7	36±2.0(12)	18.54±1.4	26.5±2.2(8.0)
	B.m + *S.m* + saw dust	21±2.0	32.2±1.9(11.2)	21.5±1.0	26± 1.5(5.5)
	B.m + *S.m* + rice husk	25±1.9	32.5± 1.1(7.5)	18±1.4	28.6±2.1 (10.6)
	B.m + *S.m* + tea waste	27.2±1.8	34.1 ± 1.8(6.9)	21±1.8	28.5±1.8(7.5)
CD (P=0.05)	(Treatments)	3.28	3.47	2.17	3.87
	(Varieties)	3.28	3.47	2.17	3.87

In vitro survivability of bacteria in bioformulations : *B. megaterium* and *S. marcescens* could survive in formulations of saw dust ($6.1\times\log_{10}$ cfu/ml), rice husk ($8.0\times\log_{10}$ cfu/ml) and in tea waste ($6.98\text{-}7.2\times\log_{10}$ cfu/ml) up to nine months, after which the decline was rapid (Fig. 1A & B).

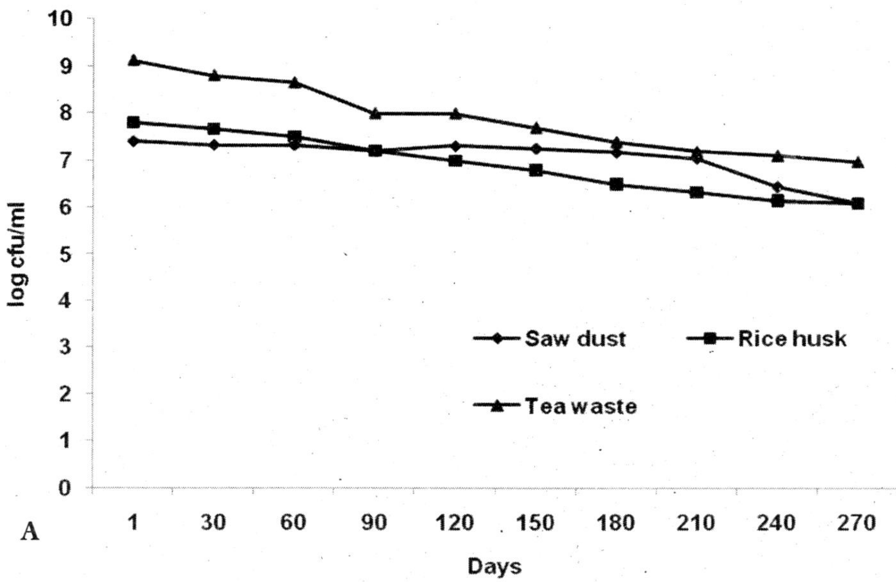

A

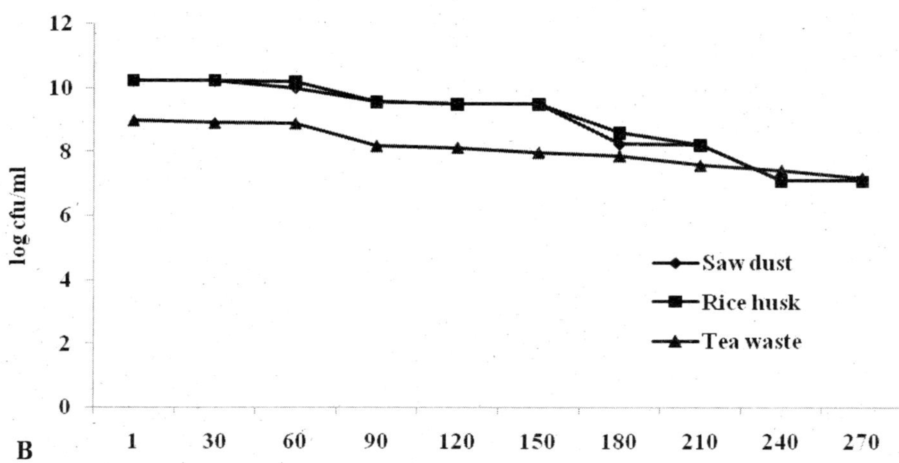

B

Figure 1 : Growth of bacteria in bioformulations at different periods of incubation. A- *B.megaterium* and B-*S.marcescens*

Discussion

Application of the bacteria as formulations is better suited for practical use. Hence, three bioformulations were prepared and initially their survivability was determined. The bacteria could survive in the bioformulations for more than nine months, after which there was in a decline stage. Previous reports are also available where *Bacillus* bioformulations could survive upto one year or more in several bioformulations (El-Hassan & Gowen, 2006). There are also several previous reports on application of PGPRs as formulations (Mathivanan *et al.* 2005; Trivedi *et al.* 2005; Hassan-El and Gowen 2006; Tilak and Reddy 2006). Due to limitations of direct inoculation and the use of various solid-phase bacterial inoculants, several polymer-based formulations, such as alginate beads, wet and dry alginate microbeads and gum-arabic preparations of bacterial species like *Azospirillum brasilense* Cd, *Pseudomonas fluorescens* and *Rhizobium sp.* have been evaluated (Bashan *et al.* 2002). Then *B. megaterium* and *S. marcescens* were also applied to the potted soil in formulated forms.. Plant growth promotion was noted and was found to be at par with the live cells, showing no significant difference. The increase in growth in the bacterial bioformulations was not due to any effect of saw dust, rice husk or tea waste, as these are neutral substances with no known plant growth promoting activities.

For commercialization, viability of bioinoculant in a prescribed formulation for a certain period with preservation of strain charactaristics is desirable (Fages 1992; Smith 1992). That's why the effects of length of storage on the viability of *B. megaterium* and *S. marcescens* in three different bioformulations are noted.

In conclusion, it may be stated that application of the bacteria in bioformulations was equally effective as aqueous suspensions. The present investigation is a step towards commercialization of the bacterial inoculants in order to reduce excessive use chemicals in tea plantations for improving its growth as well as qualities. So, the larger scale production of the bacterial inoculants in tea waste or rice husk or saw dust form is recommended.

Acknowledgement

Financial help received from Department of Biotechnology, Ministry of Science & Technology, Govt. of India is gratefully acknowledged.

References

Bashan Y, Hernandez JP, Leyva LA and Bacilio M. (2002). Alginate microbeads as inoculants carriers for plant growth- promoting bacteria. *Biol & Fertl Soil.* 35: 359-368.

Bashan, Y. (1998). Inoculants of plant growth- promoting bacteria for use in agriculture. *Biotech Adv.* 16: 729-770.

Fages J. (1992). An industrial view of *Azospirillum* inoculants: formulation and application technology. *Symbiosis.* 13: 14-22.

Glick B.R. (1995). The enhancement of plant growth by free living bacteria. Can J Microbiol. 41: 109-117.

Hassan-El SA and Gowen SR. (2006). Formulation and delivery of the bacterial antagonist *Bacillus subtilis* for management of lentil vascular wilt caused by *Fusarium oxysporum* f. sp. *lentis*. *J Phytopathol.* 154: 148-155.

Lumsden RD, Lewis JA and Fravel DR. (1995). Formulation and delivery of bio-control agents for use against soil-borne plant pathogens. In: Hall FR, Bary JW (eds), Biorational Pest Control Agents Formulation and Delivery, Washington, DC, USA, *American Chemical Society*. pp. 166-182.

Manjula K and Podile AR. (2001). Chitin- supplemented formulations improve biocontrol and plant growth promoting efficiency of *Bacillus subtilis* AF1. *Can J Microbiol.* 47: 618-625.

Mathivanan N, Prabhavathy VR, and Vijayanandraj VR. (2005). Application of talc formulations of *Pseudomonas fluorescens* Migula and *Trichoderma viride* Pers.ex S.F. Gray decrease the sheath blight disease and enhance the plant growth and yield in rice. *J Phytopathol.* 153: 697-701.

Pandey, A., Sharma, E. and Palni, and L.M.S. (1998). Influence of bacterial inoculation on maize in upland farming systems of the Sikkim Himalaya. *Soil Boil and Biochem*. 30: 379-384.

Schmidt CS, Lorenz D, Wolf GA, and Jager J. (2001). Biological control of the grapevine dieback fungus *Eutypa lata* II: Influence of formulation additives and transposon mutagenesis on the antagonistic activity of *Bacillus subtilis* and *Erwinia herbicola*. *J Phytopathol.* 149: 437-445.

Sid Ahmed A, Ezziyyani M, Sanchez CP, and Candela ME. (2003). Effect of chitin on biological control activity of *Bacillus spp* and *Trichoderma harzianum* against root rot disease in peper (*Capsicum annuum*) plants. *Eur J Plant Pathol.* 109: 633-637.

Smith RS. (1992). Legume inoculant formulation and application. *Can J Microbiol.* 38: 485-492.

Tilak KVBR and Reddy BS. (2006). *Bacillus cereus* and *B. circulans*- novel inoculants for crops. *Curr Sci.* 90: 642-644.

Trivedi P, Pandey A, and Palni LMS. (2005). Carrier-based preparations of plant growth promoting bacterial inoculants suitable for use in cooler regions. *World J Microbiol Biotech.* 21: 941-945.

Whipps, J.M. (2001). Microbial interactions and biocontrol in the rhizosphere. *J Exp Bot.* 52: 487-511.

□□□

Chapter 8

PSEUDOMONAS FLUORESCENS : A POTENTIAL BIOCONTROL AGENT FOR THE MANAGEMENT OF DAMPING OFF, FRUIT ROT AND POWDERY MILDEW IN CHILLI

R.K. MESTA, H.D. MOHANKUMAR, M. SHIVAPRASAD, M.H. TATAGAR AND P. MORE

Horticulture Research Station, Devihosur, Haveri 581 110, India

Introduction

Chilli is one of the important commercial crops of India. In India, it is grown over an area of 9.15 lakh hectares with the production of 10.18 lakh tones. Dharwad and Haveri districts, coming under northern transitional tract of Karnataka are known for export quality Byadgi chillies. However, in recent years, chilli yield is declining in this potential belt. Fungal diseases are the major constraints for the successful production of the chilli. Fruit rot caused by *Colletotrichum capsici* and powdery mildew caused by *Leivulila taurica* are the two major constraints in the production of chilli. Fruit rot can cause yield reduction to the extent of 30 per cent, besides substantial loss in the quality of the fruit incurring huge loss to the farmer. Severe powdery mildew can cause losses in yield up to 80 per cent (Gohokar and Peshney, 1981).

Hegde *et al.* (2002) tested the efficacy of 3 triazole fungicides *viz.*, hexaconazole (0.1%), propiconazole (0.1%) and triadimefon (0.1%). Among the four biocontrol agents tested *in vitro* against *C. capsici* causing leaf spot of turmeric, *Pseudomonas fluorescens* was found to be superior in inhibiting the growth of the fungus (Chidanandaswamy, 2001).

Materials and Methods

The experiment was conducted during *kharif*, 2006 and *kharif*, 2007, at Horticultural Research Station, Devihosur. The experiment on damping off was conducted in seed bed with nine treatments and three replications in Randomized Block Design. The treatments included soil treatments of *P.fluorescens* @ 2.0, 2.5 and 5 kgs /ha and and soil treatment @ 5, 10 and 20 g/kg of seeds with recommended control of carbendazim @ 2 g /kg of seeds as well as *Pythium* inoculated and un inoculated control.

The experiment for field evaluation for fruit rot and powdery mildew was laid out in Randomized Block Design with three replications and seven treatments including seedling dip and spray *Pseudomonas fluorescens* @10 g/l in various combinations with chemicals, *viz*.. carbendazim and hexaconazole. The recommended practice of 3 sprays of carbendazim and untreated control were also included in the treatment. Byadgi dabbi, a susceptible variety of chilli was used for study. The observations in seed bed were recorded on germination percentage, shoot length, root length, vigour index and percent disease incidence, were recorded for damping off and in main field on percent disease index of fruit rot and powdery mildew were recorded 15 days after the last spray, on five randomly selected plants in each treatment. The intensity of fruit rot and powdery mildew were graded by adopting 0 to 9 scale as given by Mayee and Datar (1986).

Results and Discussion

Effect of seed and soil treatments of *P.fluorescens* on germination percentage, shoot length, root length, vigour index, percent disease incidence and percent disease control have been presented in Table 1. which indicates that the difference in germination percentage of chilli seeds due to different treatments was significant. Significantly lowest percentage of germination was observed in treatment of chilli seeds with inoculated control (86.3%). All other treatments were significantly superior to this treatment and on par with each other

Shoot length differed significantly due to treatments. Seed treatment of *P. fluorescens* @ 20 g/kg of seeds (7.8 cm) recorded highest shoot length, which was significantly superior to all the treatments, but on par with seed treatment of *P.fluorescens* @ 5 g/kg of seeds (7.3 cm) and seed treatment of *P.fluorescens* @ 10 g/kg of seeds (7.0 cm). Soil application of *P.fluorescens* @ 2.5 kg / ha and 5 kg / ha were next best (both 6.8 cm). However the root length did not differ significantly due to treatments.

Vigour index differed significantly due to treatments. Highest vigour index was noticed in seed treatment of *P.fluorescens* @ 20 g/kg of seeds (1143.6) which was sigificantly superior to all the treatments. seed treatment of *P.fluorescens* @ 5 g/kg of seeds (1093.9) was next best which was on par with seed treatment of *P.fluorescens* @ 10 g/kg of seeds (1084.2). The vigour index was least in inoculated control (914.8).

Percent disease incidence (PDI) differed significantly due to treatments. Seed treatment of *P.fluorescens* @ 10 g/kg of seeds (4.59 PDI) recorded significantly least PDI than all other treatments except, seed treatment of *P.fluorescens* @ 20 g/kg of seeds (5.24 PDI) which was on par. *P.fluorescens* soil application @ 5 kg / ha was next best (7.82 PDI), which was on par with seed treatment of *P.fluorescens* @ 5g/kg of seeds (8.56 PDI). The recommended check carbendazim recorded 9.84 PDI. The un-inoculated control and inoculated controls recorded 15.26 PDI and 31.34 PDI respectively.

The per cent reduction of the disease was highest in seed treatment of *P.fluorescens* @ 10 g/kg of seeds (85.35 and 69.92 per cent) against inoculated and un inoculated controls respectively. Seed treatment of *P.fluorescens* @ 20 g/kg of seeds recorded 83.28 and 65.66 percent reduction of disease against inoculated and un inoculated controls respectively. Similarly soil application of *P.fluorescens* @ 5 kg / ha was third best with 75.05 and 48.75 percent reduction of disease against inoculated and un inoculated controls respectively.

The treatments differed significantly with respect to fruit rot and powdery mildew incidence during both 2006-07 & 2007-08 and in pooled analysis (Table 2). During 2006-07, the incidence of fruit rot was least (16.00 PDI) in the treatment 5 (seed treatment with

Table 1 : Effect of seed and soil treatments of *P.fluorescens* (SUDOCEL)on damping off in chilli (Pooled 2006-07 & 2007-08)

Sl. No.	Treatment	Germination (%)	Shoot Length (cm)	Root Length (cm)	Vigour Index	Per cent Disease Incidence
T_1	*P.fluorescens* (SUDOCEL)* soil application @ 2 kg / ha	95.5 (77.8)**	6.5 (14.8)	4.3 (12.0)	1031.4	9.12 (17.58)
T_2	*P.fluorescens* soil application @ 2.5 kg / ha	94.6 (76.6)	6.8 (15.1)	4.2 (11.8)	1040.6	9.53 (17.98)
T_3	*P.fluorescens* soil application @ 5 kg / ha	94.8 (76.8)	6.8 (15.1)	4.0 (11.4)	1023.8	7.82 (16.24)
T_4	*P.fluorescens* seed treatment @ 5 g/kg seeds	94.3 (76.2)	7.3 (15.7)	4.3 (12.0)	1093.9	8.56 (17.01)
T_5	*P.fluorescens* seed treatment @ 10 g/kg seeds	96.8 (79.7)	7.0 (15.3)	4.2 (11.8)	1084.2	4.59 (12.37)
T_6	*P.fluorescens* seed treatment @ 20 g/kg seeds	95.3 (77.5)	7.8 (16.2)	4.2 (11.8)	1143.6	5.24 (13.23)
T_7	Carbendazim 50 WP seed treatment @ 2 g / kg seeds	94.6 (76.6)	6.1 (14.3)	4.1 (11.7)	964.9	9.84 (18.28)
T_8	Inoculated (Pythium culture) control	86.3 (68.3)	6.4 (14.7)	4.2 (11.8)	914.8	31.34 (34.04)
T_9	Un inoculated control	95 (77.1)	6.3 (14.5)	4.1 (11.7)	988.0	15.26 (22.99)
	S. Em.	0.8	0.1	0.1	8.2	0.4
	CD at 5 %	2.4	0.3	NS	24.7	1.2

* SUDOCEL is commercial formulation of *P. fluorescens* supplied by Excel Crop Care Ltd Bombay

** Figures in parenthesis are angular transformed values

Carbendazim(0.2%) + seedling dip with *P.fluorescens* (10g/l)+2 Spray of *P. fluorescens* at 45,60 DAT+2 spray Hexaconazole 75 & 90DAT) which was significantly least among all other treatments. The same trend was seen during 2007-08 also. Pooled results indicated that this treatment showed significantly less percentage of fruit rot incidence (20.6) than all other treatments. During 2006-07 the incidence of powdery mildew was least (26.70 PDI) in the treatment 4 (Seed treatment with Carbendazim(0.2%) + seedling dip with *P.fluorescens* (10g/l)+2 Spray of *P. fluorescens* at 45,60 DAT+1 spray Hexaconazole

75 & 90DAT) which was significantly least than all other treatments. However during 2007-08 and in pooled results treatment 5 (28.80 PDI) (Seed treatment with Carbendazim(0.2%) + seedling dip with *P.fluorescens* (10g/l)+2 Spray of *P. fluorescens* at 45,60 DAT+2 spray Hexaconazole 75 & 90DAT) was showing significantly least percentage of powdery mildew incidence (20.60 PDI) than all other treatments. The untreated check recorded significantly highest incidence of fruit rot PDI (21.7, 38.8 & 30.3 respectively during 2006-07, 2007-08 & pooled analysis), and powdery mildew PDI (56.9, 55.5 & 53.2 respectively during 2006-07, 2007-08 & pooled analysis).

Table 2 : Management of fruit rot and powdery mildew of chilli through *Pseudomonas fluorescens*

Sl. No.	Treatments	Fruit rot (PDI)			Powdery Mildew (PDI)		
		2006-07	2007-08	Pooled	2006-07	2007-08	Pooled
1	Seed treatment (ST) with Carbendazim(0.2%) + Seedling dip (SD)with *P.fluorescens* (10g/l)+ 4 spray of *P.fluorescens* at 45,60,75,90 Days after transplanting (DAT)	18.8 (25.7)*	22.3 (28.2)	20.6 (27.0)	47.5 (43.6)	34.4 (35.9)	41.0 (39.8)
2	ST with Carbendazim(0.2%) + SD with *P.fluorescens* (10g/l) +2 Spray of *P. fluorescens* at 45,60 DAT+1 spray Carbendazim at 75 DAT	18.8 (25.7)	28.6 (32.3)	23.7 (29.1)	45.2 (42.2)	38.8 (38.5)	42.0 (40.4)
3	ST with Carbendazim(0.2%) + SD with *P.fluorescens* (10g/l)+2 Spray of *P. fluorescens* at 45,60 D AT+2 spray Carbendazim at75,90 DAT	18.2 (25.3)	25.5 (30.3)	21.9 (27.9)	42.0 (40.4)	33.3 (35.2)	37.7 (37.9)
4	ST with Carbendazim(0.2%) + SD with *P.fluorescens* (10g/l)+2 Spray of *P. fluorescens* at 45,60 DAT+1 spray Hexaconazole 75,DAT	19.6 (26.3)	24.2 (29.5)	21.9 (27.9)	26.7 (31.1)	36.6 (37.2)	31.7 (34.3)
5	ST with Carbendazim(0.2%) + SD with *P.fluorescens* (10g/l)+2 Spray of *P. fluorescens* at 45,60 DAT+2 spray Hexaconazole 75,90DAT	16.0 (23.6)	25.2 (30.1)	20.6 (27.0)	28.4 (32.2)	28.8 (32.5)	28.6 (32.3)
6	ST with Carbendazim(0.2%) + Carbendazim 3 sprays 60,75,90 DAT	19.1 (25.9)	28.9 (32.5)	24.0 (29.3)	49.0 (44.4)	42.2 (40.5)	45.6 (42.5)
7	Untreated check	21.7 (27.8)	38.8 (38.5)	30.3 (33.4)	56.9 (49.0)	55.5 (48.2)	56.2 (48.6)
	S. Em	0.31	0.80	0.61	0.32	0.71	0.42
	CD at 5 %	1.53	2.39	1.83	0.95	2.10	1.25

*Figures in parentheses are angular transformed values

Table 3: Economics of Management of fruit rot and powdery mildew of chilli

Sl. No.	Treatments	Yield (q/ha) 2006-07	2007-08	Pooled	Gross Retruns (Rs)	Gross Cost (Rs)	Net Returns (Rs)
1	Seed treatment (ST) with Carbendazim(0.2%) + Seedling dip (SD)with *P.fluorescens* (10g/l)+ 4 spray of *P.fluorescens* at 45,60,75,90 Days after transplanting (DAT)	7.0	6.8	6.9	27,600	7,680	19,920
2	ST with Carbendazi m(0.2%) + SD with *P.fluorescens* (10g/l) +2 Spray of *P. fluorescens* at 45,60 DAT+1 spray Carbendazim at 75 DAT	6.7	5.7	6.2	24,800	6,340	18,460
3	ST with Carbendazim(0.2%) + SD with *P.fluorescens* (10g/l)+2 Spray of *P. fluorescens* at 45,60 DAT+2 spray Carbendazim at75,90 DAT	7.0	6.2	6.6	26,400	6,780	19,620
4	ST with Carbendazim(0.2%) + SD with *P.fluorescens* (10g/l)+2 Spray of *P. fluorescens* at 45,60 DAT+1 spray Hexaconazole 75,DAT	8.1	6.5	7.3	29,200	6,600	22,600
5	ST with Carbendazim(0.2%) + SD with *P.fluorescens* (10g/l)+2 Spray of *P. fluorescens* at 45,60 DAT+2 spray Hexaconazole 75,90DAT	8.5	7.4	8.0	32,000	6,860	25,140
6	ST with Carbendazim(0.2%) + Carbendazim 3 sprays 60,75,90 DAT	7.4	6.3	6.9	27,600	5,660	21,940
7	Untreated check	5.4	3.9	4.7	18,800	5,000	13,800
	S. Em	0.11	0.32	0.21			
	CD at 5 %	0.32	1.01	0.60			

The treatments differed significantly with respect to yield during both the years and in pooled analysis (Table 3). During 2006-07, 2007-08 and in pooled anlysis, the treatment 5 (Seed treatment with Carbendazim(0.2%) + seedling dip with *P.fluorescens* (10g/l)+2 Spray of *P. fluorescens* at 45,60 DAT+2 spray Hexaconazole 75 & 90DAT) was significantly higher than all other treatments with 8.5 q/ha, 7.4 q/ha and 8.0 q/ha respectively. The treatment 4 which is similar to treatment 5 but lacking 1 spray of hexaconazole at 90 DAT was the next best with 8.1 q/ha, 6.5 q/ha & 7.3 q/ha respectively during 2006-07, 2007-08 and in pooled analysis. The recommended practice of seed treatment and spray of carbendazim (treatment 6) recorded 7.4 q/ha, 6.3 q/ha and 6.9 q/ha respectively, while the untreated check recorded 5.5 q/ha, 3.9 q/ha and 4.7 q/ha respectively during 2006-07,

2007-08 and in pooled analysis. The economics of the present experiment is studied in terms of net returns. The treatment 5 recorded highest net returns of Rs 25,140/-, followed by treatment 4 which recorded Rs 22,600/-.Treatment 6 (recommended practice) recorded Rs 21,940/-, while untreated check (Treatment 7) reorded net returns of Rs 25,140/-. Biological control through the use of antagonistic micro-organisms is a potential, non-chemical means of controlling plant disease by reducing inoculum levels of the pathogens. Such a management would help in preventing the pollution and also health hazards. However sole use of boiagent may not effective in the management of any plant disease as the bioagents have certain limitations. In this context in the present investigation, *Pseudomonas fluorescens* has shown its potentiality when used along with the chemicals. Seedling dip at the time of transplanting stimulates the ISR (Induced Systemic resistance) system of the plant activating the quiescent genes (Vidyasekaran *et al.*, 1997). The efficacy of *Pseudomonas fluorescens* may also be due several possibilities of existence of microbial interactions such as stimulation, inhibition, mutual intermingling of growth of antagonistic isolate over test pathogen etc. have been enumerated by many workers (Naik and Sen, 1995). Astaputhre (2006) reported that, many members of *Erysiphaceae* are parasitized by hyperparasites. The use of hexaconazole instead of the recommended carbendazim has recorded lesser incidence of powdery mildew and fruit rot and increased yield. The efficacy of hexaconazole in the management of fruit rot & powdery mildew was also previously reported by several workers. (Astaputhre *et. al.*2006, Hegde 2002).

Any management of plant disease should account for maximum profit to the farmer. In this context the in present investigation the best treatment consisting of Seed treatment with Carbendazim(0.2%) + seedling dip with *P.fluorescens* (10g/l)+2 Spray of *P. fluorescens* at 45,60 DAT+2 spray Hexaconazole 75 & 90DAT not only recorded the least disease incidence for fruit rot and powdery mildew but also gave higher net returns.

References

Astaputhre S. A. (2006). Epidemiology and management of Powdery mildew of chilli *Ph. D. Thesis*, Univ. Agric. Sci., Dharwad.

Chidanandaswamy, B. S. (2001). Studies on *Colletotrichum capsici* (Syd.) Butler and Bisby causing leaf spot of turmeric (*Curcuma longa* L.). *M.Sc.(Agri.) Thesis*, Univ. Agric. Sci., Dharwad.

Gohokar, R.T.. and Peshney, N. L.(1981). Chemical control of powdery mildew of chilli. *Indian J.Agric.Sci.*51, 663-665.

Hegde, G.M., Mesta, R.K. and Anahosur, K.H., (2002). Efficacy of triazole compounds against fruit rot pathogen (*Colletotrichum capsici*) of chilli. *Pl. Path. Newslet.*, Univ. Agric. Sci., Dharwad, 20: 3-4.

Mayee,C. D. and Datar, V. V.(1986). *Phytopathometry. Technical Bulletin-1*, Marathwada Agricultural University, Parbhani, Maharashtra, India. ; *p.29.*

Mesta, R. K. (1996). Studies on fruit rot of chilli caused by *Colletotrichum capsici* (Sydow.) Butler and Bisby. *M.Sc.(Agri.) Thesis*, Univ. Agric. Sci., Dharwad.

Naik, M. K. and Sen, B., (1995). Biocontrol of plant disease caused by *Fusarium* spp. In: *Recent Development in Biocontrol of Plant Diseases*. (Eds) Mukherjee, K.G., Tiwari,J.P.,Arora, D.K., Adithya Books Private Limited, New Delhi, p.37-51.

Vidyasekaran, P., (1987). *Fungal pathogenesis in Plants and Crops Molecular biology and Host defense Mechanisms.* p.135-146

Yarwood, C. Powdery mildews.*The Bot. Review.23,235-301.*

□□□

Chapter 9

DOMINANT RHIZOSPHERE BACTERIA AS SOURCE OF ANTIFUNGAL AGENTS AND PGPR

A. GHOSH[1], A.SAHA[2] AND D. SAHA[1]

[1]Plant Biotechnology laboratory, Department of Biotechnology, University of North Bengal, Siliguri-734013, [2]Molecular Plant Pathology and Fungal Biotechnology Laboratory, Department of Botany, University of North Bengal, Siliguri-734013

Introduction

The rhizosphere is very rich in organic compounds, supporting the growth of a variety of microbial populations, which exert beneficial, neutral or deleterious effects on plants. Beneficial soil microorganisms are those involved in improving plant growth and controlling phytopathogens by a combination of mechanisms. According to Liu *et. al.* (2007), plant growth promoting rhizobacteria (PGPR) comprises of the bacterial group that can competitively colonize plant root and enhance plant growth and/or decrease the incidence of plant diseases. Therefore, biological control of soil-borne diseases of plants involves the use of PGPR to attack and control plant pathogens and the diseases they cause. A successful biocontrol agent efficiently suppresses the pathogen and reduces disease incidence (Chet *et.al.*, 1990). Biological control can be an important component in the development of more sustainable agricultural systems because increasing use of chemical inputs causes several negative effects such as development of pathogen resistance to the applied agents and non-target environmental impacts (Compant *et.al.*, 2005).

Different PGPR traits include siderophore production, phosphate solubilization, IAA production and secretion of various hydrolytic enzymes such as chitinase, cellulase, lipase, amylase, protease etc. The siderophores are thought to facilitate biocontrol by sequestering iron from pathogens, thus limiting their growth (Leong, 1986). There are a number of reports suggesting that plant species are capable of obtaining iron from some microbial siderophores. Cline *et. al.* (1984) demonstrated that iron from microbial hydroxamate siderophores may become available to plants both in nutrient solution and in soil. Furthermore, siderophores from fluorescent pseudomonads have also been implicated in iron uptake by tomato plants (Duss *et. al.*1986).

Brinjal is an economically important crop plant in India and is frequently attacked by root pathogens such as *Fusarium equiseti* and *F. solani*. Both the pathogens cause wilt in brinjal and severe loss in crop production. Present work focuses on the isolation of different rhizospheric bacteria having antagonistic activities against these pathogens. The bacterial isolates would be studied for their *in vitro* antagonism and checked for different PGPR activities. The *in vitro* interaction of the pathogen and antagonistic bacterial strain would be studied through scanning electron microscopy (SEM).

Materials and Methods

Fungal cultures : *F. solani* (ITCC No. 4999) was obtained from Indian Agricultural Research Institute, New Delhi. *Fusarium equiseti* (ITCC Identification No. 6566.07) was earlier isolated from diseased brinjal plants. Both pathogens were maintained in potato dextrose agar (PDA) and subcultured in PDA before performing the experiments.

Isolation of bacteria : Rhizosphere soil samples were collected aseptically from different locations of brinjal cultivation fields of sub-Himalayan region of West Bengal, India. Ten gram of soil sample was kept for 2 hours on a shaker with 100 ml of sterile Winogradsky solution (Nielsen *et. al.,* 1998). Serial dilution upto 10 fold was done with distilled water and was used for spread plating on soil extract agar medium (Barrow and Feltham, 1993). The plates were incubated at 30°C for 48 hours and 104 colonies were picked and maintained as pure culture. The isolates were periodically transferred in nutrient agar slants and preserved at 4°C.

Screening for antagonistic bacterial isolates : All 104 bacterial isolates were checked for *in vitro* antagonism against *F. equiseti* and

F. solani by dual culture method described by Nielsen *et.al.* (1998). Four mm agar block taken from advancing zone of fungal hyphae growing in PDA was inoculated at the centre of each fresh PDA plate. Bacterial streaking was done in a semi-circular manner at 2-3 cm distance from the centre. A plate with only fungal block was kept as control. All the plates were incubated at 30°C until the fungal hyphae reached the edge of the control plate. Twenty isolates showed strong antagonistic activity toward both pathogens and were selected for further studies.

Siderophore production, characterization and quantification : The universal Chrome Azurol S assay (Schwyn and Neilands, 1987) was used to test the ability of bacterial isolates to produce siderophore. All 20 antagonistic strains were inoculated on CAS agar plate and incubated for 24-48 h at 30°C. Orange halo around the colony after overnight incubation indicate siderophore production (Husen, 2003). Siderophores produced were chemically characterized using previously described methods for catecholate type (Arnow, 1937) and hydroxamate type (Dave *et. al.*, 2006). A quantitative assay was done to find out siderophore units present in the culture supernatant following Machuca and Milagres (2003). The amount of siderophore present was calculated as [(Ar-As)/Ar].100 = percent siderophore units, where Ar and As are the absorbances of reference and sample at 630 nm.

Test for production of IAA : Production of indole-3-acetic acid (IAA) was determined by the method of Patten and Glick (2002). 40-45 h old bacterial cultures grown on Luria-Bertani broth supplemented with L-tryptophan was centrifuged to obtain culture supernatant. This supernatant was added with Salkowski's reagent (Gordon and Weber, 1951) and incubated at room temperature for 20-30 min. The absorbances of samples were read at 535 nm.

Test for extracellular hydrolytic enzymes

Pikovskaya's agar (Pikovskaya, 1948) was used to check phosphatase activity of the isolates. A clear halo around the colony indicates positive result. Production of lipase and amylase was tested by using Tween 80 and starch respectively as substrates in medium (Aneja, 2003). Test for pectinase and cellulase production was done following Kumar *et. al.* (2005) where M9 basal medium (Miller, 1974) was supplemented with cellulose and pectin respectively prior to inoculation. Subsequently, the plates were incubated for 5-8 days at 30^{0}C and observed for production of clear halos around the colonies, which is an indication of positive result. Chitinase activity was studied by the method of Bargabus *et. al.* (2002) with slight modification. Culture filtrates were added to the wells of petriplates containing chitinase medium (0.1% glycol chitin and 1% agarose). Plates were incubated overnight, flooded with fluorescent brightener 28 (Sigma Aldrich) and observed under UV light. No fluorescence around the well indicated positive result.

Table 1: Amount of siderophore produced by some selected antagonistic isolates

Bacterial strains	CON1	AS01	BB5	T5	CR14	CR12
% siderophore unit	73.91	43.47	39.13	21.73	26.08	13.04

SEM study

For SEM study, the cover glass carrying the interacting zone of pathogen and antagonist was collected from dual culture plates and the samples were fixed with 2.5% glutaraldehyde solution for one hour. This step was followed by dehydration through an ascending series of ethanol: 50%, 70% and 90% (vol/vol) ethanol for 5 min each; two changes with 90% ethanol for 15 min each and with 100% ethanol for 20 min each. All ethanol dilutions were made with distilled, deionized water. After dehydration, the samples were coated with gold-palladium layer (IB2-ion coater, Japan) and observed under scanning electron microscope [Model: Hitachi S-530 (Japan) 1986].

Results and Discussion

Altogether 104 strains were isolated from the rhizosphere soil of which 20 strains were found to be potent antagonists against pathogens, *F. equiseti* and *F. solani* as evident from dual culture tests (Fig. 1). Antifungal activity was found to be highest in strains AS04, AS01 and CON1 among all isolates. Earlier reports suggest that organisms present in rhizosphere soil are able to suppress plant pathogens under natural conditions (Walsh, 2001).

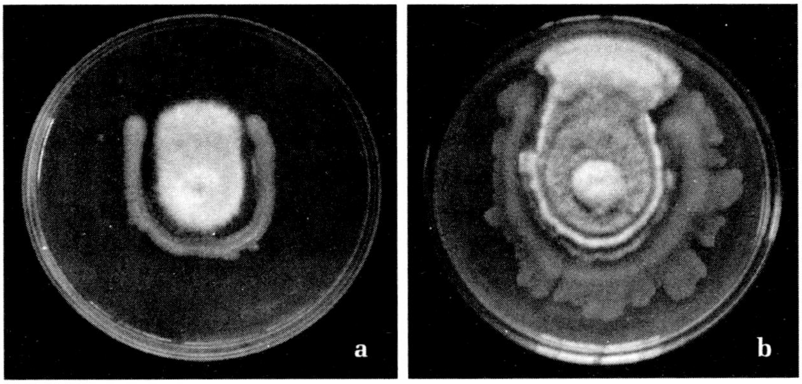

a b

Figure 1a : Suppression of *Fusarium equiseti* by AS01, **b:** suppression of *Fusarium solani* by CON1 in dual-culture test

Characterization studies revealed that all 20 antagonistic strains produced siderophore which was confirmed by orange halo around the colonies in bluish green medium (Fig. 2a). Arnow's test revealed 6 strains to produce catecholate (Fig. 2b) and Tetrazolium salt test revealed 14 isolates to produce hydroxamate siderophore (Fig.2c).

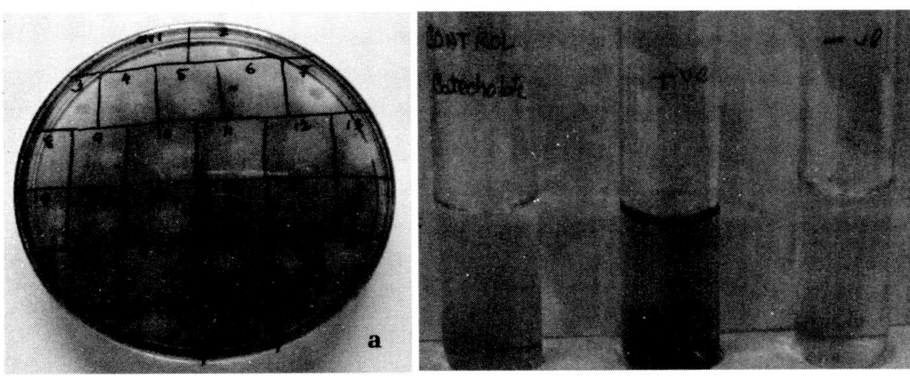

a b

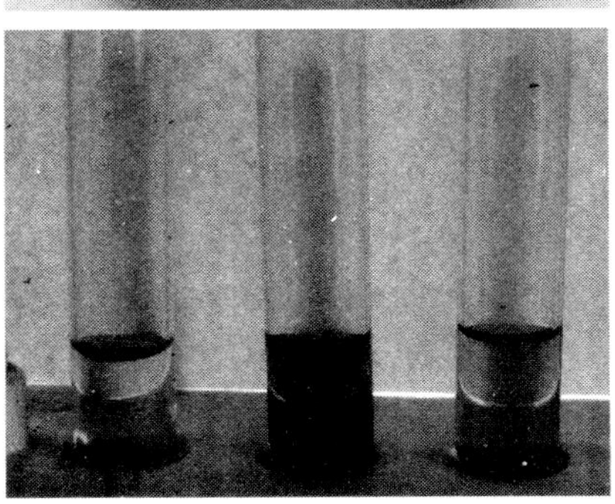

c

Figure 2a : Siderophore production on CAS agar plate, **b:** catecholate siderophore detection, **c:** hydroxamate siderophore detection

Quantification of siderophore showed CON1 to produce highest amount of siderophore (73.91% units). Amount of siderophore production ranged from 13.04% to 73.91% (Table 1). Literature reports support the fact that sequestration of iron by microbial siderophore is found to significantly increase crop yield and provide competitive inhibition to the growth of soil borne pathogens (O'Sullivan and O'Gara, 1992). The present strains produced high amounts of catecholate and hydroxamate siderophores that may be involved in depriving phytopathogens of iron.

The potent antagonists produced a number of PGPR traits other than siderophore, such as the hydrolytic enzymes: chitinase, amylase, phosphatase and lipase. Strain CON1 produced chitinase (Fig.3a), AS04 produced phosphatase and four isolates (AS01, AS04, BB5 and MB1) produced lipase (Fig.3b) and all isolates were positive for amylase production (Fig.3c). None of the isolates produced cellulase and pectinase while all produced IAA.

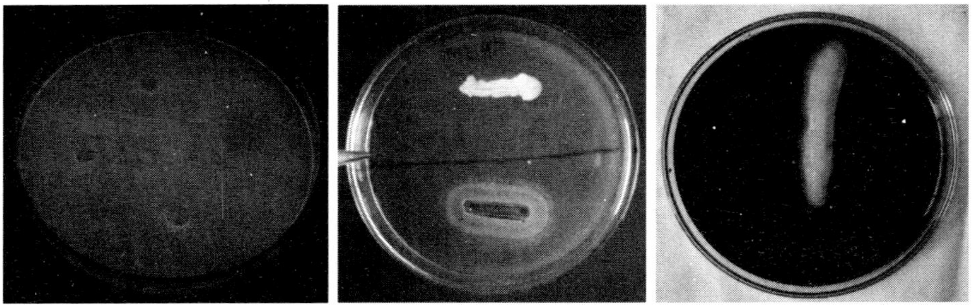

a b c

Figure 3a : Production of chitinase by CON1, **b:** lipase production, **c:** amylase production by antagonistic bacteria.

Biocontrol bacteria are known to produce different extracellular hydrolytic enzymes such as chitinase, amylase, lipase, protease etc. which aid them in being potent inhibitor of pathogens (Kishore *et. al.,* 2005). Positive results in having PGPR traits like siderophore production, IAA production, phosphate solubilization were encouraging for studying the antagonistic strains further. Chitinase production has received attention as an important property of biocontrol agents. Application of a chitinase producing strain of *Bacillus cereus* directly to soil significantly protected cotton seedlings from root rot disease caused by *Rhizoctonia solani* (Pleban *et. al.*, 1997). Cellulase and pectinase activity was not observed in any of the isolates suggesting that these enzymes are not secreted by the bacteria. Production of cellulase and pectinase is considered undesirable traits in plant beneficial bacteria (Cattelan *et. al.*, 1999) and is hence a desirable property. Additionally, all isolates were able to produce IAA. It has been documented that production of IAA by bacteria associated with plants enhances the development of host plant root system, thereby helping in growth of crop plants (Patten and Glick, 2002). Biofertilizer traits like secretion of IAA may help to increase the potential of the present isolates as bioinoculants.

SEM studies showed mycelial deformities and lysis of the hyphae (Fig. 4). Bacterial cells were found attached to the hyphal surface of *F. equiseti*. This was accompanied by clearing of the hyphal fluid in some areas. Such events may occur due to secretion of secondary metabolites and diffusible lytic enzymes by this bacterium. Several authors have reported the production of anti-fungal compounds by fluorescent pseudomonads responsible for suppression of phytopathogenic fungi (Gupta *et. al.*, 2001; O'Sullivan and O'Gara, 1992 ; Shanahan *et. al.*, 1992 ; Ge *et. al.*, 2007). Additionally some authors have suggested the involvement of secondary metabolites in morphological abnormalities when observed under microscope. Gupta *et. al.* (2001) demonstrated that the mycelial/ sclerotial abnormalities in *Macrophomina phaseolina* and *Sclerotinia sclerotiorum* occurred due to production of antifungal metabolites by the fluorescent *Pseudomonas* GRC2. Amer *et. al.* (1997) reported that deformities and lysis of the hyphae of *Fusarium oxysporum* and *Rhizoctonia solani* were caused due to attachment and penetration of antagonistic bacteria.

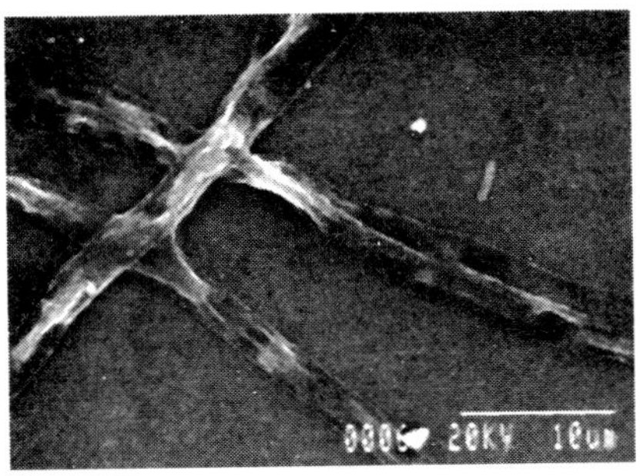

Figure 4 : SEM photography of deformed hyphae of *F.equiseti* in presence of AS01

The above results suggest that the antagonistic strains isolated during the present study may be useful in controlling plant diseases caused by *Fusarium* spp. At a time when the use of synthetic chemicals as fungicides is being curtailed, biological control agents that inhibit fungal growth may be exploited for the control of phytopathogens in the field. The present strains may thus be useful as ecofriendly fungicides after appropriate field studies.

Acknowledgement

This work was funded by University Grants Commission, Govt. of India and Deptt. of Biotechnology, Govt. of West Bengal in the form of major research projects.

References

Amer GA, Aggarwal R, Singh DV, and Srivastava KD. (1997). Interaction of *Bacillus thuringiensis* with *Pythium ultimum* and *Fusarium oxysporum f. sp. lycopersici*: possible role in biological control. *Curr Sc* 73(3):284-286

Aneja KR.(2003). Experiments in microbiology, plant pathology and biotechnology. 4th ed. New Age International (P) Ltd., New Delhi

Arnow LE. (1937). Colorimetric Determination of the Compounds of 3,4-Dihydroxyphenylalanine- tyrosine Mixture. *J Biol Chem* 118: 531-537

Bargabus RL, Zidack NK, Sherwood JE and Jacobsen BJ. (2002). Characterization of systemic resistance in sugarbeet elicited by a non pathogenic, phyllosphere-colonizing *Bacillus mycoides*, biological control agent. *Physiol Mol Pl Pathol* 61:289-298

Barrow GI and Feltham RKA. (1993). Cowan and Steel's Manual for the Identification of Medical Bacteria, 3rd ed. Cambridge University Press

Cattelan, AJ, Hartel PG, and Fuhrmann JJ. (1999). Screening for Plant Growth–Promoting Rhizobacteria to Promote Early Soybean Growth. *Soil Sci. Soc. Am. J.* 63: 1670-1680

Chet I, Ordentlich A, Shapira R, and Oppenheim A. (1990). Mechanisms of Biocontrol of Soil-Borne Plant Pathogens by Rhizobacteria. *Plant and soil* 129: 85-92

Cline G.R, Reid C.P.P, Powell P.E. and Szaniszlo P.J. (1984). Effects of a Hydroxamate Siderophore on Iron Absorption by Sunflower & Sorghum. *Plant Physiol* 76: 36-39

Compant S, Duffy B, Nowak J, Clement C and Barka EA. (2005). Use of plant Growth-Promoting Bacteria for Biocontrol of Plant diseases: Principles, Mechanisms of Action, and Future Prospects. *Appl and Environ Microbiol* 71(9): 4951-4959

Dave BP, Anshuman K and Hajela P. (2006). Siderophores of Halophilic Archaea and their Chemical characterization. *Indian J Exp Biol* 44: 340-344

Duss F, Moazfar A, Oertii JJ and Jaeggi W. (1986). Effect of Bacteria on the Iron Uptake by Axenically-Cultured-Roots of Fe-efficient and Fe-inefficient Tomatoes (*Lycopersicon esculentum* Mill). *J Plant Nutr* 9:587-598

Ge YH, Pei DL, Zhao YH, Li WW, Wang SF and Xu YQ. (2007). Correlation between antifungal agent phenazine-1-carboxylic acid and pyoluteorin biosynthesis in *Pseudomonas* sp. M18. *Curr Microbiol* 54: 277-281

Gordon SA and Weber RP. (1951). Colorimetric estimation of Indoleacetic Acid. *Plant Physiol* 26: 192-195

Husen E. (2003). Screening of Soil Bacteria for Plant Growth Promotion Activities *in vitro*. *Indonesian J Agri Sc* 4 (1): 27-31

Gupta CP, Dubey RC, Kang SC and Maheshwari DK. (2001). Antibiosis-mediated necrotrophic effect of *Pseudomonas* GRC2 against two fungal plant pathogens. *Curr sc* 81 (1): 91-94

Kishore GK, Pande S and Podile AR. (2005). Biological Control of Late Leaf Spot of Peanut (*Arachis hypogaea*) With Chitinolytic Bacteria. *Biol Control* 95: 1157-1165

Kumar RS, Ayyadurai N, Pandiaraja P, Reddy AV, Venkateswarlu Y, Prakash O and Sakthivel N. (2005).Characterization of antifungal metabolite produced by a new strain *Pseudomonas aeruginosa* PUPa3 that exhibits broad-spectrum antifungal activity and biofertilizing traits. *J Appl Microbiol* 98: 145-154

Leong J. (1986). Siderophores: their biochemistry and Possible Role in the Biocontrol of Plant Pathogens. *Annu Rev Phytopathol* 24: 187-209

Liu H, He Y, Jiang H, Peng H, Huang X, Zhang X, Thomashow LS and Xu Y. (2007). Characterization of a Phenazine-Producing Pseudomonas chlororaphis GP72 with Broad- Spectrum Antifungal Activity from Green Pepper Rhizosphere. *Curr Microbiol* 54: 302-306

Machuca M and Milagres AMF. (2003). Use of CAS Agar Plate Modified to Study the Effect of Different Variables on the Siderophore Production by *Aspergillus*. *Letters in Appl Microbiol* 36: 177-181

Miller JH. (1974). Experiments in Molecular Genetics, 2nd ed. *Cold Spring Harbor Laboratory Press*, Cold Spring Harbor, NY

Nielsen MN, Sorensen J, Fels J, and Pedersen HC. (1998). Secondary Metabolite and Endo-Chitinase-Dependent Antagonism Toward Plant Pathogenic

Microfungi of *Pseudomonas fluorescens* Isolated from Sugarbeet Rhizosphere. *Appl Environ Microbiol* 64: 3563-3569

O'Sullivan J and O'Gara F. (1992). Traits of Fluorescent *Pseudomonas* spp. Involved in Suppression of Plant Root Pathogens. *Microbial Review* 56: 662-676

Patten CL and Glick BR. (2002). Role of *Pseudomonas putida* Indoleacetic Acid in Development of the Host plant Root System. *Appl Environ Microbiol* 68: 3795-3801

Pikovskaya RI. (1948). Mobilization of Phosphorus in soil in Connection with Vital Activity of Some Microbial Species. *Mikrobiologiya* 17: 363-370

Pleban S, Chernin L and Chet I. (1997). Chitinolytic activity of an endophytic strain of *Bacillus cereus*.*Letts Appl Microbiol* 25: 284-288

Schwyn B, and Neilands JB. (1987). Universal Chemical Assay for the Detection and Determination of Siderophore. *Annual Biochem* 180: 47-56

Shanahan P, O'Sullivan DJ, Simpson P, Glennon JD and O'Gara F. (1992). Isolation and characterization of an antibiotic-like compound from a fluorescent pseudomonad and investigation of physiological parameters influencing its production. *Appl Environ Biol* 58: 353-358

Walsh UF, Morrissey JP and O'Gara F. (2001). *Pseudomonas* for Biocontrol of Phytopathogens: from Functional Genomics to Commercial Exploitation. *Curr Opinion in Biotech* 12: 289–295

□□□

Chapter 10

EVALUATION OF *TALAROMYCES FLAVUS* FOR PHOSPHATE SOLUBILIZATION AND BIOCONTROL ACTIVITY AND ITS MOLECULAR ANALYSIS

K. SUNAR, P.L. DEY, U. CHAKRABORTY AND B.N. CHAKRABORTY

Immuno–Phytopathology Laboratory, Department of Botany,
University of North Bengal, Siliguri 734013, Darjeeling, West Bengal, India

Introduction

Phosphorus (P), one of the major plant nutrients whose availability is conditioned by various factors, is acquired from soil solution as phosphate anions (predominantly HPO_4^{2-} and $H_2PO_4^-$), and these anions are extremely reactive and may be immobilized through precipitation. The availability of phosphate anions in soil is attributed to the presence and activity of soil microorganisms. The existence of soil microorganisms capable of transforming soil P to forms available to the plants has earlier been reviewed (Wansik *et.al.*, 2006). Among the various PSF isolates obtained from soil, most efficient phosphorus solubilizers are species of *Aspergillus* (Chakraborty *et al.*, 2008). Strains of *Aspergillus niger* and *Penicillium* are the most common fungi capable of phosphate solubilization. Many bacterial, fungal, yeast, and actinomycetes species capable of solubilizing sparingly soluble phosphorus in pure culture have been isolated and studied (Halder *et al.*, 1991; Abd-Alla,1994; Whitelaw, 2000; Goldstein, 1986). *Talaromyces flavus* (anamorph, *Penicillium dangeardii*; sometimes reported as *Penicillium vermiculatum*) has shown ability to convert insoluble phosphorus (P) to an accessible form, like orthophosphate, which is an important trait for increasing plant yields. Taxonomy of *Talaromyces flavus* is currently based largely on morphological character such as conidial form, size, color and ornamentation, branching pattern with short side. Identification of *Talaromyces flavus* isolates at the species level has proved difficult, due to the degree of morphological similarities with various species of *Penicillum*. It was found that morphological analysis was highly prone to error and roughly 50% of the identification of *Talaromyces flavus*.

Molecular methods have recently been introduced into *Talaromyces flavus* taxonomy with revision. These techniques have been proven to be valuable tools in fungal taxonomy and their application has led to the reconsideration of several genera (Sherriff *et al.,* 1994). In *Talaromyces flavus,* isoenzyme analysis and rDNA sequencing have been used to distinguish species within specific groups of strains.

The purpose of the present study was establishment of *Talaromyces flavus* gene sequence in North Bengal isolates based on the sequence analysis of ITS-1 regions of the rDNA gene for rapid identification.

Materials and Methods

Isolation and identification of Talaromyces flavus : Soil samples were collected from rhizophere soil of plantation crops and agricultural fields of North Bengal. The location of soil samples were recorded through GIS mapping tool (Garmin). *Talaromyces flavus* species were isolated in specific selective medium which contained $MgSO_4.7H_2O$ -0.2; K_2HPO_4-0.9; KCl-0.15; NH_4NO_3-1.0; glucose-3.0, chloramphenicol-0.25, fenaminosulf-0.3, pentachloronitrobenzene-0.2, rose Bengal-0.15, captan-0.02 (post autoclaving), agar-20 (g/lit);

Screening for phosphate solubilizing activity : Screening for primary phosphate soliblizing activity of the isolate was carried out by allowing the fungi to grow in selective media, i.e., Pikovskaya's agar (Pikovskaya, 1948) medium for 7 to 10 days at 25^{0}C. The appearance of a transparent halo zone around the fungal colony indicated the phosphate soliblizing activity of the fungus.

Estimation of phosphorus : Phosphorus solubilizing ability of fungal strains was tested in two different types of Pikovskaya (PVK) liquid medium (Pikovskaya, 1948) amended with with 0.5% tricalcium phosphate and 0.25% (w/v) Rock phosphate (RP-140), pH of the medium was adjusted to 7 before autoclaving. Dissolved phosphate concentration in the culture filtrate was determined by ammonium molybdate-ascorbic acid method (Kundsen and Beegle, 1988). Difference between the amount of tricalcium phosphate and rock phosphate added and their remainder after incubation gave the exact phosphate solubilizing potential of the isolates.

In vivo testing of the isolate on growth of different crops : Effect on growth of fungus amended with soil was conducted using *Oryza sativa*, *Glycine max*, *Phaseolus vulgaris* and *Vigna radiata* in non sterile pot conditions arranged in a completely randomized design. Surface sterilized seeds were sown in pre amended soil with mycelia and spores in pots (18 cm dia). After germination and a few days of growth, soil phosphate concentration and growth of plants were recorded after intervals of 10, 20 and 30 days.

Antagonistic test of *T. flavus in vitro*

Dual plate culture: *T. flavus* was tested for its *in vitro* antifungal activity against *S. rolfsii* by dual culture technique. The fungal pathogen was obtained from Immuno-Phytopathology Laboratory, Department of Botany, University of North Bengal and were maintained with regular sub culturing in PDA for subsequent tests. Both the test organism and the pathogen were grown separately in the petriplates and inocula were cut from the growing region and placed in fresh sterile PDA plates. In each plate, inoculum block of the isolate and of the test pathogen were placed 4 cm apart on the agar medium. Interactions between *T. flavus* and *S. rolfsii* were observed at different intervals from 4th day onwards and antagonism or inhibition of the pathogen was determined.

***In vivo* antagonistic test of *T. flavus* against *S. rolfsii* :** Antagonistic test against *S. rolfsii* under green house condition was conducted following the method of Madi *et al.* (1997) with modifications. Seeds of *Vigna radiata* were surface disinfested with 0.1% $HgCl_2$ for two minutes, rinsed three times with sterile water and immersed in spore suspension (10^6/ml) of *T. flavus* for 24 h. Seeds were also similarly treated with spore suspension of *Trichoderma* *harzianum*, a known biocontrol agent which served as a positive check and seeds soaked in sterile distilled water served as control. Plastic pots (10 cm dia) were filled with 500 g of sterilized garden soil. Ten surface sterilized bean seeds were placed on the soil surface of each pot; 3-4 sclerotia of *S. rolfsii* were placed 0.5–1 cm apart from each bean seed. Seeds were then covered with 150 g of sterile soil, and the pots were incubated in the green house at temperatures ranging from 28 to 30°C. Seedling emergence and survival percentage were recorded for 30 days at an interval of 5 days after planting. There were four treatments in the experiment: (a) distilled water (control); (b) S. *rolfsii* (c) *S. rolfsii* + *T. flavus* and (d) *S. rolfsii* and *T. harzianum* (positive control) The experiment was conducted with four replicates per treatment where each pot served as a replicate, and the data was pooled for analysis.

Genomic DNA extraction from isolates of Talaromyces flavus : Isolation of fungal genomic DNA was done. Fungal mycelia from 3-4 days old culture was crushed with liquid nitrogen and incubated with lysis buffer containing 250 mM Tris-HCl (pH 8.0), 50 mM EDTA (pH8.0), 100 mM NaCl and 2% SDS, for 1 hr at 65°C followed by centrifugation at 12,000 rpm for 15 min. The supernatant was extracted with equal volume of water saturated phenol, centrifuged at 12,000 rpm for 15 min, and further extracted with equal volume of phenol : chloroform : isoamyl alcohol (25:24:1) by centrifugation at 12000 rpm for 15 min; the aqueous phase was transferred in a fresh tube and chloroform (in the ration of 1:4 v/v) was added followed by 0.5M Na-acetate (in the ration of 1:10 v/v). Next, isopropanol was added to the above mixture (0.7 times the final volume) and centrifuged. DNA was precipitated from the aqueous phase with chilled ethanol (100%) and pelleted by centrifuging at 12000 rpm for 15 min followed by washing in 70% ethanol and centrifugation. The pellets were air dried and suspended in 1xTE buffer pH 8.

Qualitative and quantitative estimation of DNA : The extraction of total genomic DNA from the *Talaromyces flavus,* as per the above procedure was followed by RNAase treatment. Genomic DNA was re suspended in 100 μl 1 X TE buffer and incubated at 37°C for 30 min with RNAse (60 mg). After incubation the sample was re-extracted with PCI (Phenol: Chloroform: Isoamylalcohol (25:24:1) solution and

RNA free DNA was precipitated with chilled ethanol as described earlier. The quality and quantity of DNA was analyzed both spectrophotometrically and in 0.8% agarose gel. The DNA from all isolates produced clear sharp bands, indicating good quality of DNA.

PCR amplification of ITS region of Talaromyces flavus isolates: All isolates of *Talaromyces flavus* were taken up for ITS-PCR amplification. Genomic DNA was amplified by mixing the template DNA (50 ng), with the polymerase reaction buffer, dNTP mix, primers and Taq polymerase. Polymerase Chain Reaction was performed in a total volume of 100 μl, containing 78 μl deionized water, 10 ml 10 X Taq polymerase buffer, 1ml of 1U Taq polymerase enzyme, 6 μl 2 mM dNTPs, 1.5 μl of 100 mM reverse and forward primers and 1 μl of 50ng template DNA. For amplification of the ITS1–5.8S–ITS2 region of *Talaromyces flavus* isolates, the primer pair T/ITS1 TCTGTAGGTGAACCTGCGG and T/ITS4 TCCTCCGCTTAT TGATATGC was used.

PCR was programmed with an initial denaturing at 94°C for 5 min. followed by 30 cycles of denaturation at 94°C for 30 sec, annealing at 59°C for 30 sec and extension at 70°C for 2 min and the final extension at 72°C for 7 min in a Primus 96 advanced gradient Thermocycler. PCR product (20 μl) was mixed with loading buffer (8 μl) containing 0.25% bromophenol blue, 40 % w/v sucrose in water, and then loaded in 2% Agarose gel with 0.1 % ethidium bromide for examination with horizontal electrophoresis. The strain of *Talaromyces flavus* produced a single amplified product ranging from 800—850bp and PCR products were sent for sequencing to Genie, Bangalore.

Data analysis : The sequenced PCR product was aligned with nine ex-type strain sequences from NCBI GenBank and established fungal taxonomy for identification. Sequences were aligned following the Clustal Walgorithm (Thompson *et al.*, 1994) included in the Megalign module (DNASTAR Inc.). Multiple alignment parameters used were gap penalty = 10 and gap length penalty = 10. Both of these values are aimed to prevent lengthy or excessive numbers of gaps. The default parameters were used for the pairwise alignment. The use of Clustal W determines that, once a gap is inserted, it can only be removed by editing. Therefore, final alignment adjustments were made manually in order to remove artificial gaps. Phylogenetic analyses were completed using the MEGA package (version 4.01; Institute of Molecular Evolutionary Genetics, University Park, PA). Neither gaps (due to insertion-deletion events) nor equivocal sites were considered phylogenetically informative. Hence, complete deletion prevented the use of any of these sites in further analyses. Phylogenetic inference was performed by the UPGMA method. Bootstrap tests with 1,000 replications were conducted to examine the reliability of the interior branches and the validity of the trees obtained. An additional standard error test was performed with the data set using the same characters in order to evaluate the statistical confidence of the inferred phylogeny.

Results and Discussions

Soil inhabiting fungus *Talaromyces flavus* (RHS/P 51) was isolated from the paddy growing soil. Microscopic observation revealed that the fungus was filamenteous and possessed white to yellow ascoma composed of loose hyphae, ascospores yellow thick walled subglobose to ovoid 3.5 – 4 .0 x 2.7 – 3.5 μm (Fig 1). The fungus produces colonies reaching upto 5-7 cm dia in 7 days at 25-30°C of incubation on PDA. The results and the descriptions were in accordance to the description by Malloch and Cain (1973), where the species is described as anamorphs mostly belonging to *Penicillium.*

Phosphate solubilization properties of the isolate was detected on Pikovskaya's agar medium (Pikovskaya, 1948) where the formation of halos around the colony gave a positive result after incubation for 7 to 10 days at 25°C. Further, the utilization of Tricalcium phosphate and rock phosphate under *in vitro* conditions was calculated at an interval of 7 days which was found to be higher in comparison to the other native phosphate solubilizing fungi like *Aspergillus niger* (Table 1). Species of *A. niger* are reported to be seed borne pathogens (Chakraborty and De, 2008) and is a common soil inhabiting fungus (Gochenaur, 1967). Owing to their pathogenic character they were not considered much for further plant growth promotion studies but however was taken up for comparative analysis. In our present investigation *T. flavus* solubilized 849 mg/l of tricalcium and 374 mg/l of rockphosphate respectively after ten days of incubation.

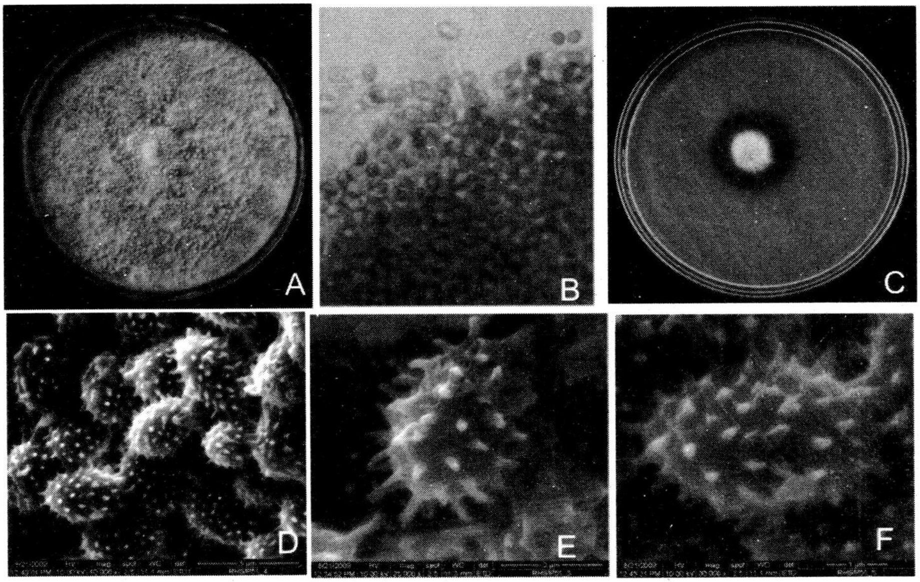

Figure 1 : Growth of *Talaromyces flavus* in PDA medium (A), Spores in bright field (B), Phosphate solubilization activity of *T. flavus* on PKV medium, SEM of *T. flavus* (D-F)

Table 1 : Evaluation of Phosphate solubilizing potential of PSF isolates [*T. flavus* and *A. niger*] in liquid medium ammended with tricalcium phosphate and rock phosphate

	Tricalcium Phosphate (mg/g)		Final pH	Rock Phosphate (mg/g)		Final pH
	7 days	14 days		7 days	14 days	
T. flavus	512	849	3.5	175	374	3.0
A. niger	488	756	3.0	112	243	3.0

Root and shoot lengths of the potted plants grown in soil amended with *T. flavus* were found to be more in comparison to other amendments . Results revealed that the fungus could enhance plant growth in pot conditions compared to control plants grown in soil without any amendments (Table 2).

The growth promotion was due to increase in soil microbial activity where bio availability of P was greatly enhanced. Soil phosphate content and growth of certain cereals was enhanced when soil was inoculated with fungus prior to seed sowing and germination.

Table 2 : Effect of soil amendment with *T. flavus* on plant growth

Plants	days	Control	Amended with *T. flavus*
Phaseolus vulgaris	10	2.15 ± 0.7	2.30 ± 0.7
	20	12.7 ± 0.3	13.4 ± 0.6
	30	16.2 ± 0.8	18.5± 0.9
Glycine max	10	3.28 ± 0.6	3.21 ± 0.6
	20	11.9 ± 0.3	13.3 ±0.5
	30	16.0 ± 0.6	21.9 ± 1.1
Oryza sativa	10	12.1 ± 0.2	12.3 ± 0.2
	20	17.45 ± 0.7	21.9 ± 1.1
	30	20.1 ± 1.3	24.7 ± 1.6
Vigna radiata	10	11.1 ± 0.1	14.3 ± 0.3
	20	16.45 ± 0.7	20.9 ± 1.2
	30	21.1 ± 1.7	27.7 ± 1.7

Values are average of 30 plants; ± Standard Error

T. flavus could inhibit mycelial growth and development of *S. rolfsii* in dual culture as well as a number of other fungal pathogens *in vitro* (Table 3).

Table 3 : Inhibition of plant pathogenic fungi by *T. flavus* in dual culture

Test fungi	Isolates obtained from	Percent inhibition by *T. flavus**
Alternaria alternata	*Camellia sinensis*	51.55
Bipolaris sorokiana	*Triticum aestivum*	60.45
Curvularia lunata	*Camellia sinensis*	55.75
Fusarium oxysporum	*Lycopersicon esculentum*	71.33
Fusarium solani	*Citrus reticulata*	75.55
Macrophomina phaseolina	*Glycine max*	62.45
Magnaporthe grisea	*Oryza sativa*	71.75
Rhizoctonia solani	*Triticum aestivum*	73.75
Sclerotiana sclerotiorum	*Camellia sinensis*	71.66
Sclerotium rolfsii	*Vigna radiata*	85.50
Sphaerostilbe repens	*Camellia sinensis*	70.00

*Cultures grown on PDA as dual culture at 28°C for 12 days

The efficacy of *T. flavus* as a biocontrol agent against *S. rolfsii* was further confirmed *in vivo*, since it inhibited seedling emergence and survival of *Vigna radiata* when tested in the green house. Highest inhibition (70%) of seedling survival against *S. rolfsii* was evident following treatment with ascospore suspension of *T. flavus* even after 30 days of planting (Fig. 2.), which confirmed the biocontrol potential of *T. flavus*, where isolates of *T. flavus* suppressed stem rot in mungbean caused by *S.rolfsii*. *T. flavus* has also been reported to suppress stem rot of mungbean caused by *S. rolfsii* (Dethoup *et al.*, 2007) *Verticillium* wilt of tomato and eggplant (Fahima and Henis, 1992; Madi *et al.*, 1997; Marois *et al.*, 1984; Tjamos and Fravel, 1997) and parasitize *Sclerotinia sclerotiorum* (Huang *et al.*, 2000) and *Sclerotium rolfsii* (Madi *et al.*, 1997). The suggested mechanisms of biocontrol against plant pathogens include mycoparasitism (McLaren *et al.*, 1986; Madi *et al.*, 1997), antibiotic production (Stosz *et al.*, 1996) and competition (Marois *et al.*, 1984). In addition, *T. flavus* has also been reported to produce several cell wall degrading enzymes responsible for antagonistic activity against phytopathogenic fungi (Madi *et al.*, 1997). *T. flavus* chitinase inhibited cell wall formation in

Figure 2 : Effects of *Talaromyces flavus* in plant health improvement
A&E-*Oryza sativa*, B&F-*Phaseolus vulgaris* C&G-*Glycine max* D&H-*Vigna radiata*
(A-D) Control, (E-H) Treated with *T. flavus*

Verticillium dahliae, *Sclerotinia sclerotiorum* and *Rhizoctonia solani*, and also inhibited spore germination and germ tube elongation of *Alternaria alternata*, *Fusarium moniliforme* and *Magnaporthe grisea* (Duo-Chuan *et al.*, 2005). Nagtzaam and Bollen (1997) reported colonization of roots of eggplant and potato by *T. flavus* from coated seed. They found that the ability of *T. flavus* to colonize plant roots may contribute to disease suppression by reducing the proliferation of the pathogens on the roots by direct mycoparasitism or competition (Table 3 and Fig. 3).

Table 4 : Identified *Talaromyces flavus* and comparison with referred NCBI GenBank

Strain No	GenBank accession no	Identified as	Identity (%)	Reference
RHS/P 51	GU324073	*Talaromyces flavus*	100	Chakraborty *et al.*, 2010
NRRL 2098	EU021694	*Talaromyces flavus*	100	Peterson, S.W. 2010
NRRL 2098	EU021663	*Talaromyces flavus*	100	Peterson, S.W. 2010
NRRL 2098	EU021620	*Talaromyces flavus*	100	Peterson, S.W. 2010
CBS 310.38	FJ530982	*Talaromyces flavus*	100	Varga, J. *et al.*, 2010
ZE0199	GQ204181	*Talaromyces flavus*	100	Jumpathong *et al.*, 2009
South-west0092	FJ537107	*Talaromyces flavus*	100	Zhao, Z. *et al.*, 2009
CBS22566	AY766252	*Talaromyces flavus*	100	Luangsa-ard, J.J. 2004
FRR 2386	L76925	*Talaromyces flavus*	100	Geiser, D.M. 1996
18354	U18354	*Talaromyces flavus*	100	Berbee, M.L. 1995

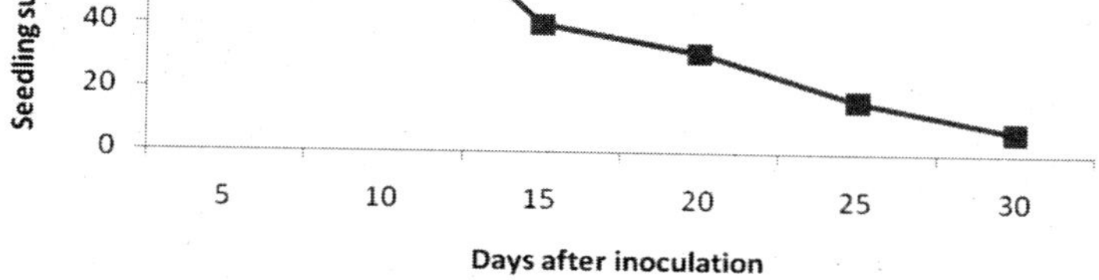

Figure 3 : Percentage survival of mungbean seedlings following treatment of seeds with *T. flavus* and inoculation with *S. rolfsii* . Tf=*Talaromyces flavus*, Sr=*Sclerotium rolfsii*. Average of 4 replicate pots per treatment with 10 seedlings per pot.

The ribosomal RNA genes (rDNA) possess characteristics that are suitable for the identification of fungal isolates at the species level. These rDNA are highly stable and exhibit a mosaic of conserved and diverse regions within the genome (Hibbett, 1992). They also occur in multiple copies with up to 200 copies per haploid genome (Bruns, 1991; Yao, 1992) arranged in tandem repeats with each repeat consisting of the 18S small subunit (SSU), the 5.8S, and the 28S large subunit (LSU) genes. Internal transcribed spacer (ITS) regions have been used successfully to generate specific primers capable of differentiating closely related fungal species (Bryan, 1995). In the broader context, taxon-selective amplification of ITS regions is likely to become a common approach in molecular identification strategies, In the present study, we focused on the ITS regions of ribosomal genes for the construction of primers that can be used to identify *Talaromyces flavus* ITS region of rDNA was amplified using genus specific ITS-1 and ITS4 primers. Amplified products of size in the range of 800bp was produced by the primers. Following direct sequencing of the PCR products of *Talaromyces flavus* sequences were obtained that could be aligned and showed satisfactory homology with nine ex-type strains of *Talaromyces flavus* sequences from the NCBI Genbank data base as analysed by BLAST (Table 4). The priming site of the ITS1 and ITS4 primers were determined in order to confirm that the sequences obtained corresponding to the actual ITS 4 region. A multiple sequence alignment was carried out that included the ITS 1 region, including gaps and the complete sequences align. There were quite a number of gaps that were introduced in the multiple sequence alignment within the ITS-4 region that were closely related and similar sequence indicated whether *Talaromyces flavus* isolates were used in the multiple sequence alignment. From the sequence alignment, variations were observed between other nine ex-type *T. flavus* isolates. The ITS PCR has helped to detect polymorphism at ITS region of rDNA among the *Talaromyces flavus*. On the first approach all the conserved regions of 18S r DNA sequences with other ex-type isolate sequences obtained from NCBI Gene bank data base were analyzed using the bioinformatics tool BioEdit. A multiple sequence alignment was carried out that included the ITS region, including gaps and the complete sequences align. There were quite a number of gaps that were introduced in the multiple sequence alignment within the region that were closely related and similar sequence indicated whether the isolates were closely related (Fig. 4). The evolutionary history was inferred using the UPGMA method (Sneath & Sokal, 1973). The optimal tree with the sum of branch length = 2.79809851 is shown. The percentage of replicate trees in which the associated taxa clustered together in the bootstrap test (1000 replicates) are shown next to the branches (Felsenstein, 1985). The tree is drawn to scale, with branch lengths in the same units as those of the evolutionary distances used to infer the phylogenetic tree. The evolutionary distances were computed using the Maximum Composite (99%) Likelihood method (Tamura *et al.*, 2004) and are in the units of the number of transversional substitutions per site. Codon positions included were 1st+2nd+3rd+Noncoding.

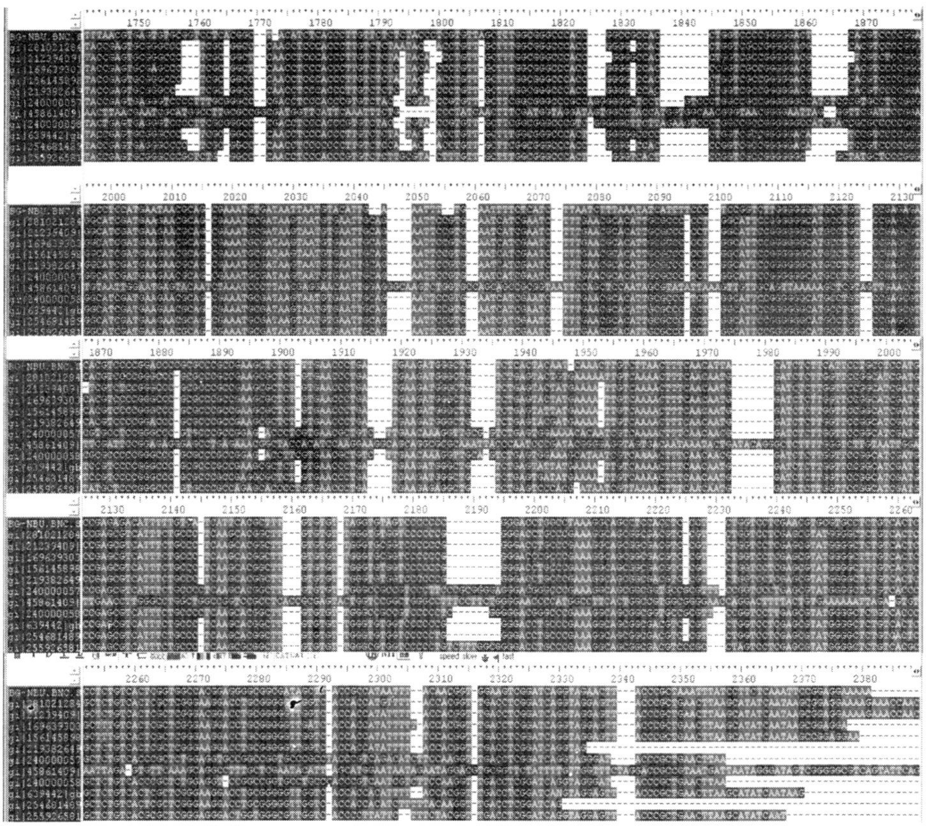

Figure 4 : 18S rDNA sequence alignments of *Talaromyces flavus*. Data for other species were gathered from NCBI. The conserved regions of the gene are demonstrated in different colour

All positions containing gaps and missing data were eliminated from the dataset (Complete deletion option). There were a total of 138 positions in the final dataset. Phylogenetic analyses were conducted in MEGA 4 as described by Tamura *et al.*, 2007) and is presented in Fig. 5.

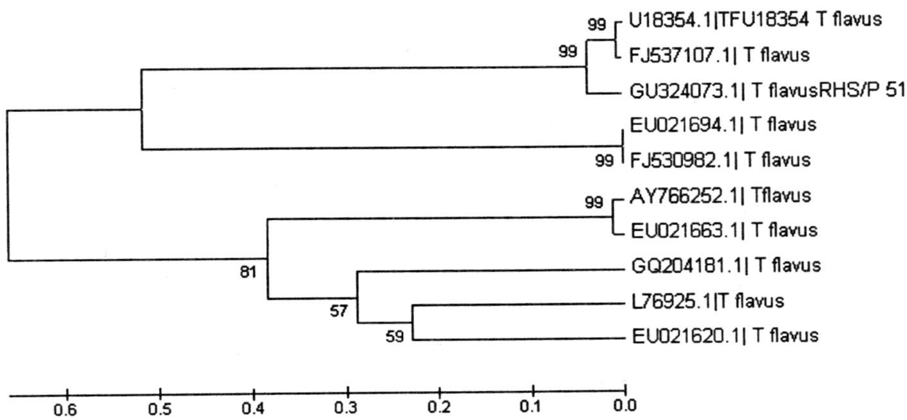

Figure 5 : The phylogenetic analyses conducted using the UPGMA method among the isolate of *T. flavus* (RHS/P51) with other ex-type isolates of *T. flavus* by MEGA4.1 software.

In conclusion, above results indicated that the species of *Talaromyces* under study could be identified as *T. flavus* on the basis of available similar "r DNA-based sequence", of *T. flavus*.

Acknowledgement

Financial support from ICAR, as a part of network project entitled Application of Microorganisms in Agriculture and Allied Sectors (AMAAS) of National Bureau of Agriculturally Important Microorganisms (NBAIM), is gratefully acknowledged.

References

Abd Alla, M.H. (1994). Phosphatases and the utilization of organic phosphorus by *Rhizobium leguminosarum* biovar *viceae*. *Lett. Appl. Microbiol.* 18 : 294–296.

Bruns T.D., White T.J. and Talyor J.W., (1991). Fungal molecular systematics. *Annu. Rev. Ecol. Sys.* 22: 525-564.

Bryan G.T., Daniels M.J. and Osbourn A.E. (1995). Comparison of fungi within the *Gaeumannomyces-Phialophora* complex by analysis of ribosomal DNA sequence. *Appl. Environ. Microbiol,* 61: 681-689

Chakraborty, B.N., Chakraborty, U., Saha, A., Dey, P.L. and Sunar, K. (2008) Screening of phosphate solubilizing *Aspergilli* isolates from soils of North Bengal and their effects on soybean *J Mycol Plant Pathol* 38(2): 227-233.

Chakraborty, U. De, P and Chakraborty, B. (2008). Immunodetection of Asprgillus flavusin stored seed of cowpea, lentil and mugbean. *J Mycol Pl.Pathol* 38(2): 216-220.

Dethoup, T., Manoch, L., Visarathanonth, N., Chamswarng, C. and Kijjoa, A. (2007). Morphology and distribution of *Talaromyces flavus* from soil and potential use as a biological control agent against plant pathogenic fungi. *Thai J. Agric. Sc.* 40: 37-50.

Duo-Chuan, L.I., Chen, S.and Jing, L.U. (2005). Purification and partial characterization of two chitinases from the mycoparasitic fungus *Talaromyces flavus. Mycopathologia.* 159:223-229.

Fahima, T., Madi, L.and Henis, Y. (1992). Ultrastructure and germinability of *Verticillium dahliae* microsclerotia parasitized by *Talaromyces flavus* on agar medium and in treated soil. *Biocontrol Science and Technology.* 2: 69-78.

Felsenstein J (1985) Confidence limits on phylogenies: An approach using the bootstrap. *Evolution* 39:783-791.

Gochenaur, S.E. & Whittingham, W.F. (1967). Mycology of willow and cottonwood lowland communities in southern Wisconsin. I. Soil Microfungi in the willow-cottonwood forests. *Mycopath.Mycol.Appl* 33, 125-139.

Goldstein, A.H. (1986). Bacterial solubilization of mineral phosphates: historical perspectives and future prospects. *Am. J. Altern. Agric* 1 :51-57.

Halder, A.K., Mishra, A.K. and Chakarbarthy, P.K. (1991). Solubilization of inorganic phosphate by Bradyrhizobium. *Ind. J. Exp. Biol.* 29: 28-31.

Hibbett D.S., (1992). Ribosomal RNA and fungal systematics. *Trans. Mycol. Soc.,* 33: 533-556.

Huang, H.C., Bremer E., Hynes, R.K. and Erickson, R.S. (2000). Foliar application of fungal biocontrol agents for the control of white mold of dry bean caused by *Sclerotinia sclerotiorum. Biol. Cont.* 18:270-276.

Kundsen, D. and Beegle, D. (1988). Recommended phosphorous tests. P.122-15, In : W.C.Dahnke(ed.) Recommended chemical soil tests procedures for the north central region. Bulletin no. 499 (revised) North Dakota agric.exp.sta.,Fargo, North Dakota.

Madi, L., Katan, T., Katan, J. and Henis, Y. (1997). Biological Control of *Sclerotium rolfsii* and *Verticillium dahliae* by *Talaromyces flavus* is mediated by different mechanisms. *Phytopathology.* 87: 1054-1560.

Malloch, D. & Cain, R.F.,(1973). The Trichocomataceae, Ascomycetes with *Aspergillus Pacelomyces* and *Penicellium* imperfect states. *Can J. Bot,* 50, 2613-2628

Marois, J.J., Fravel, D.R. and Papavizas, G.C. (1984). Ability of *Talaromyces flavus* to occupy the rhizosphere and its interaction with *Verticillium dahliae. Soil Biol. Biochem.* 16: 387-390.

Mclaren, D.L., Huang, H.C. and Rimmer, S.R. (1986). Hyperparasitism of *Sclerotinia sclerotiorum* by *Talaromyces flavus. Can. J. Plant Pathol.* 8: 43-48.

Nagtzaam, M.P.M. and Bollen, G.J. (1997). Colonization of roots of eggplant and potato by *Talaromyces flavus* from coated seed. *Soil Biol. Biochem.* 29:1499-1507.

Pikovskaya, R.I. (1948). Mobilization of phosphorus in soil connection with the vital activity of some microbial species. *Microbiologia* 17 :362–370.

Sherriff, C., M.J. Whelan, G.M. Arnold, J.F. Lafai,Y. Brygoo and J.A. Bailey, (1994). Ribosomal DNA sequence analysis reveals new species groupings in the genus *Colletotrichum. Exp Mycology,* 18: 121–138.

Sneath PHA & Sokal RR (1973) Numerical Taxonomy. Freeman, San Francisco.

Stosz, S.K., Fravel, D.R. and Roberts, D.P. (1996). *In vitro* analysis of the role of glucose oxidase from *Talaromyces flavus* in biological of the plant pathogen *Verticillium dahliae. Appl Environ. Microbiol.* 62:3183-3186.

Tamura K, Dudley J, Nei M and Kumar S (2007) MEGA4: Molecular Evolutionary Genetics Analysis (MEGA) software version 4.0. *Mol Biol Evolution* 24: 1596-1599.

Tamura K, Nei M & Kumar S (2004) Prospects for inferring very large phylogenies by using the neighbor-joining method. *Proceedings of the National Academy of Sciences* (USA) 101:11030-11035.

Thompson, J., D. Higgins and T. Gibson, (1994).CLUSTAL W: Improving the sensitivity of progressive multiple sequence alignment through sequence weighting, position-specific gap penalties and weight matrix choice. *Nucleic Acid Research* 22: 4673-4680.

Tjamos, E.C. and Fravel, D.R. (1997). Distribution and establishment of the biocontrol fungus *Talaromyces flavus* in soil and on roots of solanaceous crops. *Crop. Prot.* 16:135-139.

Wansik, S. Jeounghyun, R. Yongjoo, K. Jinchul, Y. Munusamy, M. and Tongmin, S. (2006). Phosphate Solubilization and Growth Promotion of Maize (*Zea mays* L.) by the Rhizosphere Soil Fungus *Penicillium oxalicum*. *18th World Congress of Soil Science* 140-8.

Whitelaw, M.A. (2000). Growth promotion of plants inoculated with phosphate solubilizing fungi. (Ed. In: by Donald L. Sparks. *Advances in Agronomy*, Academic press 69 : 99-151.

Yao C, Frederiksen R.A., Magill C.W., (1992). Length heterogeneity in ITS2 and the methylation status of CCGG and GCGC sites in the rRNA genes of the genus Peronosclerospora. *Curr. Genet.* 22: 415-420.

□□□

Chapter 11

SELECTION OF BIOANTAGONISTIC *PSEUDOMONAS FLOURESCENS* FOR THE BIOLOGICAL CONTROL OF *FUSARIUM OXYSPORUM* F. SP. *LYCOPERSICI* CAUSING WILT OF TOMATO

C. SRINIVAS AND D. NIRMALA DEVI

Department of Microbiology & Biotechnology, Bangalore University, Janna Bharathi Campus, Bangalore-560 056, India Ph: 080-22961461

Introduction

Fusarium wilt caused by the fungal pathogen *Fusarium oxysporum* Schlechtend: Fr. f. sp. *lycopersici* (Sacc.) W.C. Snyder & H.N. Hans is known as one of the most devastating diseases of tomato worldwide (Walker J.C. 1971). In India, most of the varieties are susceptible and fungicides are frequently used to control the disease (Chauhan, 1988; Kapoor.1988). However, the fungicide applied at the time of sowing may not persist for the whole cropping season. In this context, biological control is an alternative strategy for disease management. Several microorganisms have been successfully used for the control of tomato wilt (Podile *et al.*, 1985; De Cal *et al.*, 1995; Padmodaya and Reddy, 1998). In recent years, there has been much success in obtaining effective control of plant pathogens using beneficial biocontrol agents such as strains of *Pseudomonas* species which have been used extensively for plant growth promotion and disease control, because of many properties such as efficient colonization of under ground plant organs, utilization of a large number of organic substrates commonly found in root exudates and production of a variety of secondary toxic to fungi and bacteria. They are also inexpensive to grow and can be introduced into the rhizosphere by seed treatments or as granules (Srivatsava *et al.*, 2001).

The main objective of the present study is to isolate antagonistic *P. flourescens* which has antagonistic effect on *F. oxysporum in vivo* and *in vitro* and their characterization in terms of antagonistic mechanisms to control the pathogen, and to assess the suitable conditions for growth similar to those present in the field.

Materials and Methods

Isolation, selection, innoquity and identification of bacteria : Different strains of *P. fluorescens* were isolated from the rhizosphere and rhizoplane of tomato from farmer's field in Karnataka, India. After removing the loosely adhering soil from freshly excised roots, root segments (1g) were shaken in 100 ml of sterile distilled water for one hour, diluted to 10^{-3} to 10^{-7} for selection (Vidhyasekaran and Muthamilan, 1995). *P. fluorescens* was isolated and maintained using King's medium B (KMB) (King *et al.*, 1954). The identity of the organism was further confirmed by conducting various biochemical tests specific to *P. fluorescens* (Stanier *et al.*, 1966).

The isolated strains of *P. flourescens* were tested against *F. oxysporum*. An aliquot (20μl) of suspension (5x10^9cfumL^{-1}) of 24hrs old cultures of *P. flourescens* were placed on 0.5cm sterile paper disks. The disks were placed in the centre of the Petri dish containing PDA, surrounded by a 10mm diameter of mycelial disc of a four day old culture of F. oxysporum. It was incubated for 48hrs at 22°C, and inhibition of fungal growth was checked. The strains which showed bioantagonistic activity were selected and maintained by regular sub culturing.

In order to test the pathogencity of selected *P. flourescens* to tomato, plants (two-three leaf stage) were treated immersing previously wounded (sterile needle) roots of tomato seedlings of PKM variety in a) a suspension of the antagonistic bacteria (5$\times$$10^9$cfu mL^{-1} in 2% methyl cellulose at p^H 7.0) for 60 seconds (Raupach and Kleopper, 1998), b) sterile distilled water(control-1): c) 2% methyl cellulose at p^H 7.0 (Control-2). Once treated, seedlings were placed in pots containing a mixture of perlite : vermiculite =1:1(w/w). Seedlings were maintained under green house at 15-25°C with daily watering with the addition of fertilizers, root dry weight and damage were checked after plants reached a development corresponding to four-five true leaves or 30 days.

When pathogenecity was tested on seeds, these were immersed in bacterial solution (5×10^9cfu mL^{-1}in 2% methyl cellulose at p^H 7.0) for five minute (Raupach and Kleopper, 1998). Seeds were allowed to germinate and treated in the same conditions as for seedlings but evaluation was done when seedlings were at two to three leaf stage. This included the same parameter plus germinating days.

Each experiment was repeated six times and data was subjected to arcsine transformation and analysis of variance (ANOVA) (SPSS. Version16). Significant effects of treatments were determined by the F values (Pd" 0.05) using Tukey's HSD test.

Antagonism

Dual cultures: Pure culture of *F. oxysporum* (10mm disk) was placed at the centre of a Petri dish containing PDA. A circular line made with a 6cm diameter dipped in a suspension of bioantagonistic bacteria (5×10^9cfumL^{-1}) was placed surrounding the fungal inoculum. Plates were incubated for 72hrs at 22°C and growth parameter of the pathogen was measured. Percent inhibition was calculated using the following formula: %inhibition= 1-(Fungal growth/control growth) $\times$100, Student test was performed as Pd" 0.05.

Production of diffusible antibiotics: Plates with PDA medium covered with a cellophane membrane, were inoculated in the centre with 100μl of a bioantagonistic *P. flourescens* suspension (5×10^9cfu mL^{-1}). After incubation for 22hrs at 22°C, the membrane with bacterial growth was removed and the plate was inoculated in the middle with a 10mm disk of a pure culture of *F. oxysporum*. Plates were further incubated at 22°C for 48hrs and then growth of the pathogen was measured. Control plates were kept with sterile distilled water.

Production of volatile antibiotics: The bioantagonistic *P. flourescens* suspension (100μl) (5×10^9 cfumL^{-1}) was placed at the centre of one half petridish containing king B medium, and a10-mm disk of four days old pure culture of *F. oxysporum* was placed at the centre of another petridish containing PDA. Both half plates were placed face to face preventing any contact between the pathogen and the *P. flourescens* suspension and were sealed to isolate the inside atmosphere and to prevent loss of volatiles formed. Plates were incubated at 22°C for 48hrs and the growth of the pathogen was measured and compared to control developed in the absence of selected *P. flourescens*.

Secretion of enzymes involved in bioantagonism: *P. flourescens* was cultured either on plates with solid medium or in tubes with liquid medium after inoculation with 100 μl of a suspension (5×10^9cfumL^{-1}). Media contained either chitin or yeast glucans as the sole carbon source. Development of *P. flourescens* was assessed through colony counting after incubation at 22°C for 48hrs for the plates with solid media or through measurement of absorbance at 600nm in liquid medium at 22°C for 48hrs. Also concentrated supernatants from liquid cultures of each *P. flourescens* were analyzed for the presence of endochitinase 1-3 glucanase and protease activity, as described by (Zaldivar *et al.*, 2001 and Perez *et al.*, 2002).

Effect of Fe^{+3} on antagonism level: This effect was tested according to Pumarino, 1995 using $FeCl_3\times6H_20$ at 0, 5, 10, 15 and 20 parts per million added to the King B medium.

Effect of different salt concentrations, p^H values and temperatures on development of antagonistic bacteria: The growth of bioantagonistic *P. flourescens* was tested under NaCl concentrations of 0, 50,100,150,200 and 250mM; at p^H values which ranged between 5.0 and 8.0 and temperatures of incubation of 5,10,15,28,and 37°C, 0.5mL of a *P. flourescens* suspension (5×10^9cfu mL$^{-1)}$ from a 24 hrs old culture were placed in a tube containing 4.5mL of peptone broth previously adjusted at the desired NaCl concentration and /or p^H value. Tubes were incubated for 48hrs at the temperatures mentioned above, and bacterial growth was estimated at 600nm, after serial dilutions.

Bacterial ability for root colonization: Eight pre germinated tomato seeds were placed on Petri dishes containing water agar, and the ability of bacteria to colonize roots was established according to Misaghi 1990. For evaluation, radicle apex was isolated and placed in tubes containing $MgSO_4$, subjected to Ultrasound 50-55khz for five minutes and the bacterial suspension obtained was diluted. An aliquot (0.1ml) of the diluted suspension was seeded on petridish containing

King B medium and incubated for 24hrs at 28°C and cfu were counted. The experimental unit was one petridish with eight pre-germinated seeds.

Results & Discussion

Bacterial antagonists: Twenty isolates of *P. flourescens* were initially isolated (Table1) and from these 3 isolates which showed maximum antagonistic properties against *F. oxysporum* and designated as BS.1, BS.5 and BS.18 and were selected for further study experiments.

Antagonistic mechanisms

Direct antagonism through Dual cultures: No contact was observed with any of the *P. flourescens* tested and an inhibitory halo was observed suggesting the presence of fungi was static because of metabolites secreted by the bacteria. On the other hand, a change in mycelial colour was observed closed to the edge of the colony of *F. oxysporum*, being darker brown than the one observed at the centre of the colony. This effect observed could be due to the secretion of antibiotics by *P. flourescens*, because no enzymes involved in biocontrol, such as chitinases, and/or glucanases and/or protease, could be detected as described by Silva *et al.* (2001). When *F. oxysporum* was tested against the three bacterial antagonists, maximum inhibiting effect was observed with the isolate of B.S.1 (Table 2)

Diffusible and Volatile antibiotics: Isolate BS.1 showed maximum inhibitory effect on the growth of *F. oxyporum* isolates tested, although other two isolates of *P. flourescens* showed inhibitory effect on the growth of *F. oxysporum* (Table 3 & 4)

Secretion of enzymes involved in bioantagonism: None of the three bacteria was able to grow in solid or in liquid media that contained chitin or yeast glucans as the sole carbon source. Endochitinase or 1-3 glucanase or protease activity was not detected in supernatants from liquid cultures suggesting that enzymes, if synthesized, are not secreted, or not synthesized at all by the *P. flourescens*. If we consider that none of three bacteria was able to secrete any of the enzymes involved in biocontrol, and that all had the ability to control *F. oxysporum* through secretion of diffusible and volatile metabolites, it may be concluded that they use these two latter mechanisms of biocontrol as opposite to some fungal biocontrol microorganisms that also use fungal cell wall hydrolyzing enzymes within their biocontrol mechanisms (Perez *et al.*, 2002).

Effect of Fe^{3+} on antagonism: Iron is a fundamental element for respiration of several aerobic and facultative anaerobic microorganisms, and therefore, its availability in soil is essential (Leon, 1986). On the otherhand siderophores are low weight compounds with high affinity for Fe^{3+} (Neilands 1981) which are produced under limiting concentrations of iron. These compounds are able to transport this element inside the cell for metabolic functions. (Press *et al.*, 2001), and microorganisms which are able to produce siderophores show competitive advantages as compared to those that donot produce them. From this point of view, the competence for iron increases in conditions where this element is limiting but this condition is reverted when iron is added to the culture medium (Elad & Baker, 1985). The three antagonistic bacteria showed similar behavior on *F. oxysporum* growth at any of the Fe^{3+}concentratons tested **(Table 5).** If it is considered that iron available in soils fluctuate between 1.8 and 27ppm it may be concluded that this element is not limiting for the antagonistic activity of these bacteria on *F. oxysporum*.

Bacterial growth at different NaCl concentrations: The growth of three *P. flourescens* isolates tested as antagonists of *F. oxysporum* under different concentrations of NaCl showed isolate BS.1 best growth at 50,100, 150 and 200 mM NaCl. Less growth was observed at 250mM NaCl concentration.BS.5 and BS.18 showed best growth at 50, 100NaCl concentrations. These results confirms that factor high concentrations of NaCl are not limiting for the growth of any of the bacteria (Fig. 1)

Bacterial growth at different p^H values: Isolates BS.1 and BS.5 grew best at p^H values 5, 6 and 7 and showed its highest growth at p^H 6 and 7, BS.1 showed its highest growth at p^H6 and the three *P. flourescens* did not grow at p^H9.0 (Fig. 2).These results agree in part with those which describe the optimum growth of Pseudomonas is between 5.7 and 6.8 (Claus and Berkley, 1986). In addition, it is important to mention that p^H values where the highest growth of these isolates was obtained are coincident with the optimum p^H for tomato development which is between p^H5 and 7.0. On the other hand, soils from where bacteria were obtained showed a p^H value between 6.2 and 7.2 when suspended in water suggesting that soil p^H does not affect their growth.

Table 1 : Pre evaluation of antagonistic properties of *P. fluorescens* isolated from rhizosphere & rhizoplane of healthy tomato plants, collected at different tomato growing fields of Karnataka.

Potential antagonistic *P. fluorescens* isolates	*F. oxysporum* growth (mm)	% Inhibition of *F. oxysporum* radial growth
Control	3.5 ± 0.058^{a}	0.00 ± 1.65^{g}
BS1	2.2 ± 0.058^{g}	37.1 ± 1.65^{a}
BS2	3.1 ± 0.058^{bcd}	11.43 ± 1.64^{def}
BS3	3.5 ± 0.058^{a}	0.00 ± 1.65^{g}
BS4	2.8 ± 0.058^{def}	20.00 ± 1.65^{bcd}
BS5	2.7 ± 0.058^{ef}	22.86 ± 1.65^{cb}
BS6	3.2 ± 0.058^{abc}	8.58 ± 1.65^{efg}
BS7	2.8 ± 0.058^{def}	20.00 ± 1.65^{bcd}
BS8	3.5 ± 0.058^{a}	0.00 ± 1.65^{g}
BS9	3.0 ± 0.058^{bde}	14.28 ± 1.65^{bde}
BS10	2.9 ± 0.058^{cdef}	17.15 ± 1.65^{cdef}
BS11	2.9 ± 0.058^{cdef}	17.15 ± 1.65^{cdef}
BS12	3.3 ± 0.058^{ab}	5.72 ± 1.65^{gf}
BS13	3.0 ± 0.058^{bde}	14.28 ± 1.65^{cdef}
BS14	2.8 ± 0.058^{def}	20.00 ± 1.65^{bcd}
BS15	3.1 ± 0.058^{bcd}	11.43 ± 1.65^{def}
BS16	2.9 ± 0.058^{cdef}	$17.15\pm1.^{65bcde}$
BS17	3.5 ± 0.058^{a}	0.00 ± 1.65^{g}
BS18	2.6 ± 0.058^{f}	25.72 ± 1.65^{b}
BS19	2.9 ± 0.058^{def}	17.15 ± 1.65^{bce}
BS20	3.2 ± 0.058^{abc}	8.58 ± 1.65^{gf}

Bacterial growth at different temperatures: The lowest development for all three isolates was observed at 5°C and 10°C and little growth observed at 15°C and isolate BS.1 and BS.18 shows better growth at 28°C and no growth at 40°C, while isolate BS.1 and BS.5 grew better at 37°C. (Fig. 3). When considering that soil temperatures are close to 30°C at the production season (Gonzalez), bacterial growth will not be affected by this condition. This fact, along with results obtained for growth at different Fe^{3+}or NaCl concentrations and different p^H values, suggest that development of three isolates should not be affected under the field conditions.

Bacterial capacity to colonize tomato roots: The use of *P. flourescens* to exert an appropriate biological control of *F. oxysporum* and other soil borne fungi relies on their ability to colonize roots efficiently. The ability to colonize roots is highly variable between rhizobacteria, being this characteristic a reflection for their ability to compete for ecological niches in the rhizosphere (Misaghi, 1990).

The results obtained from the present study it may be concluded that the biocontrol effect of antagonistic bacteria isolated from soils (*P. fluorescens* BS.1, BS.5 and BS.18) against *F. oxysporum* are adequate for their use at the field level, especially BS.1 isolate. The mechanisms used by *P. fluorescens* are the secretions of volatile and diffusible metabolites. In addition, they show growth characteristics, which are compatible to the conditions of salinity, p^H, Fe^{3+}, and temperature that can be found in soils where tomato is commonly cultivated. Therefore they could be used at the field level to biocontrol the *F. oxysporum* disease in tomato.

Growth of *F. oxysporum* was evaluated after 72 hours at 22°C in PDA medium.

The Experiment was performed three times and values are the mean with in the column sharing the same letters are not significantly different according to Tukey's HSD at pd" 0.05.

Table 2 : Growth inhibition of *F. oxysporum* in dual cultures against *P. fluorescens* isolates.

Isolates	**Fungal growth(mm)**	**% Inhibition**
BS.1	2.2 ± 0.5^{b}	37.14 ± 1.6^{a}
BS.5	2.5 ± 0.5^{a}	28.57 ± 1.6^{b}
BS.18	2.3 ± 0.5^{ab}	34.28 ± 1.6^{ab}

Growth of *F. oxysporum* was evaluated after 72hours at 22°C in PDA medium.

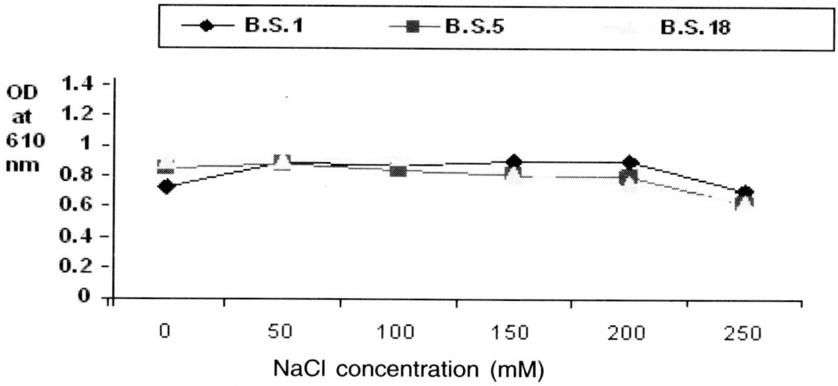

Figure 1 : Growth of *P. fluorescens* at different NaCl concentrations

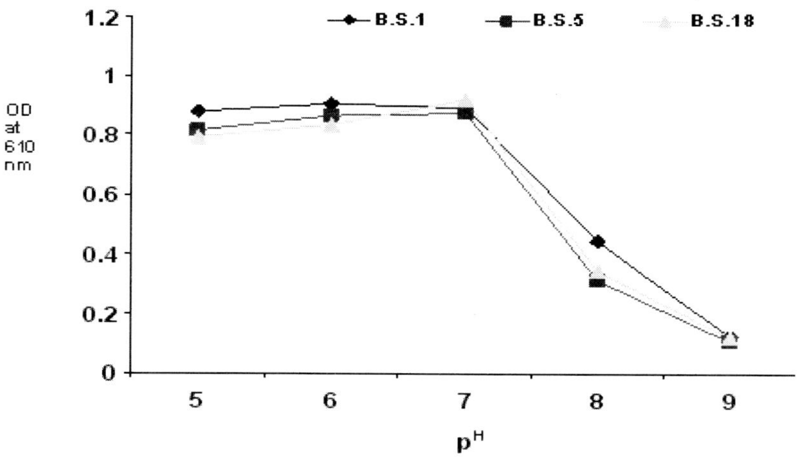

Figure 2 : Growth of *P. fluorescens* isolates at different p^H values.

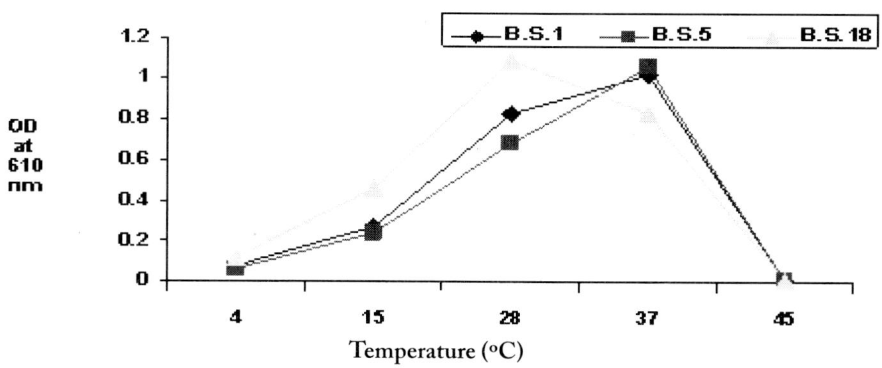

Figure 3 : Growth of *P. fluorescens* isolates at different temperatures

The Experiment was performed three times and values are the mean with in the column sharing the same letters are not significantly different according to Tukey's HSD at pd" 0.05.

Table 3 : Effect of diffusible antibiotics secreted by *P. fluorescens* isolates on radial growth of *F. oxysporum*

Isolates	Fungal growth(mm)	% Inhibition
BS.1	2.8±0.05[a]	31.42±1.6[a]
BS.5	3.1±0.05[b]	20.71±2.3[b]
BS.18	3.0±0.05[b]	28.56±1.6[ab]

Growth of *F. oxysporum* was evaluated after 72 hours at 22°C in PDA medium.

The Experiment was performed three times and values are the mean with in the column sharing the same letters are not significantly different according to Tukey's HSD at pd" 0.05.

Table 4 : Effect of volatile antibiotics secreted by *P. fluorescens* on radial growth of *F. oxysporum*

Isolates	Fungal growth (mm)	%Inhibition
BS.1	2.4±0.05[b]	31.42±1.64[a]
BS.5	3.0 ±0.05[a]	15.24±2.51[b]
BS.18	2.7±0.05[b]	22.6±4.29[ab]

Growth of *F. oxysporum* was evaluated after 72 hours at 22°C in PDA medium.

The Experiment was performed three times and values are the mean with in the column sharing the same letters are not significantly different according to Tukey's HSD at pd" 0.05.

Growth of *F. oxysporum* was evaluated after 72 hours at 22°C in PDA medium.

The Experiment was performed three times and values are the mean with in the column sharing the same letters are not significantly different according to Tukey's HSD at pd" 0.05.

Table 5 : Effect of different concentrations of Fe^{3+} on the radial growth of *F. oxysporum* in the presence of *P. fluorescens*

Fe^{3+} as $FeCl_3$ (µM)	BS1		BS5		BS18	
	Growth of *F. oxysporum* (mm)	% Inhibition	Growth of *F. oxysporum* (mm)	% Inhibition	Growth of *F. oxysporum* (mm)	% Inhibition
0	1.9±0.05^{bc}	45±1.73^{ab}	2.0±0.05^{b}	42.3±1.45^{a}	2.1±0.05^{b}	40±1.4^{a}
5	2.1±0.05^{ab}	40±1.45^{bc}	2.0±0.05^{b}	42.3±1.45^{a}	1.9±0.05^{b}.	45±1.7^{a}
10	2.2±0.05^{a}	37±1.73^{c}	2.4±0.05^{a}	30.6±1.76^{b}	2.6±0.05^{a}	26±1.1^{b}
15	1.8±0.05^{c}	48±1.73^{a}	1.9±0.05^{b}	45±1.73^{a}	2.1±0.05^{b}	40±1.4^{a}
20	2.3±0.05^{a}	34±1.73^{c}	2.1±0.05^{b}	39.6±1.45^{a}	2.4±0.05^{a}	30±1.7^{b}

References

Baker, K, and Cook, R. (1974). Biological control of plant pathogens. W. H. Freeman Company, San Francisco, USA, 433p.ISBN o835772306.

Beckman, C.H. (1987). The nature of wilt diseases of plants. *The American Phytopathological Society. St. Paul.174pp.*

Chauhan, M.S. (1988). Reaction of varieties/lines of tomato to wilt. *Indian J.Mycol.Pl.Path.*18:72-73.

De Cal, A., S.Pascual, I. Harena and P.Malgarejo. (1995). Biological control of *Fusarium oxysporum* f.sp.*lycopersici. Pl.Path.*44:909-917.

Elad,Y and Baker. R. (1985). Influence of trace amounts of cations and siderophore-producing pseudomonads on chlamydospore germination of Fusarium oxysporum. *Phytopathology*, 57: 1047-1052.

Gonzalez, R. (2001). Control biologico de Fusarium solani(Mart) Sacc. en Lycopersicon esculentum Mill mediante bacterias y Trichoderma spp. Tesis (Memoria de Titulo Ingenieria Agraria). Santiago, chile, Universidad de chile, Facultad de Ciencias Agrarias y Forestales, 70p.

Kapoor, I.J. (1988). Fungi involved in tomato wilt syndrome in Delhi, Maharashtra and Tamil Nadu. *Indian Phytopath.* 41:208-213.

Leong, J. Siderophores (1986). Their biochemistry and possible role in the biocontrol of plant pathogens. *Annual Review of Phytopathology*, 24: 187-209.

Padmodaya, B. and H.R.Reddy. (1998). Screening of antagonists against *Fusarium oxysporum f. sp.lycopercisi* causing seedling disease and wilt in tomato. *Indian J.Mycol.Pl.Pathol.*28:339-341.

Podile, A.R.,.Prasad G.S and Dube H.C. (1995). Bacillus subtilis as antagonist to vascular wilt pathogens. *Current Sci.*54:864-865.

Press C.M., Loper J.P. and Kleopper J.W. (2001). Role of iron in Rhizobacteria-mediated induced systemic resistance of cucumber. *Phytopathology*, 91(6): 593-598.

Neilands, J.B. (1981). Microbial iron compounds. Annual Review of Biochemistry., 50: 715-731.

Misaghi,J. (1990). Screening bacteria for root colonizing ability by a rapid method. Soil Biology Biochemistry, 22: 1085-1088.

Perez, L.M.,Besoain, X., Reyes, M., Pardo, G. and Montealegre,J. (2002). The expression of extra cellular fungal cell wall hydrolytic enzymes in different *Trichoderma harzianum* isolates correlate with their ability to control *Pyrenochaeta lycopersici*. Biological Research, 35(3-4): 401-410.

Pumarino, A. (1995). Evaluation in vitro del control biologicode la Fusariosis del rejol. Tesis (Memoria de Titulo Ingenieria Agaria). Santiago, Chile, Universidad de Chile, Facultad de Ciencias Agarias y Forestales. 70p.

Raupach, G.S. and Kleopper,J.W. (1998). Mixtures of plants growth-promoting rhizobacteria enhance biological control of multiple cucumber pathogens. *Phytopathology*, 88(11): 1158-1164.

Silva.p.; Montealegre, J.; Besoain X. and Perez, L.M. (2001). Effecto de bacterias biocontroladoras sobre germination y desarrollo de tomate y sobre expression de proteinas de defense. In: *Congreso Sociedad Chilena de Fitopatologia*. (XI, 4th-5th December, 2001, Santa Cruz, Chile). Abstracts, p.29.

Vidhyasekaran, P., and Muthamilan, M. (1995). Development of formulations of *Pseudomonas fluorescens* for control of chickpea wilt. *Plant Dis.* 79: 782-786.

Walker, J.C. (1971). Fusarium Wilt of Tomato. Monograph6. The American Phytopathological Society, St. Paul. MN.

Zaldivar, M.; Velasquez, J.C.; Contresas, I. and Perez L.M. (2001). *Trichoderma aureoviride* 7-121, a mutant with enhanced production of lytic enzymes: its potential use in waste cellulose degradation and/or biocontrol. *Electronic Journal of Biotechnology* [online]. 15 December, vol.4, no. 3. Available from Internet: http://www.ejbiotechnology.info/content/vol4/issue3/full/7/index.html.

□□□

Chapter 12

RHIZOSPHERE MYCOFLORA OF *CITRUS RETICULATA* AND THEIR DIVERSITY ANALYSIS

B.N. CHAKRABORTY, U. CHAKRABORTY, S. ALLAY, S. KHATI, U.DE AND P.L. DEY

Immuno-Phytopathology Laboratory, Department of Botany, University of North Bengal, Siliguri 734013, West Bengal, India

Introduction

The loose skinned mandarin orange (*Citrus reticulata*) is a traditional horticultural crop of Darjeeling district and is mostly grown as homested gardens by the villagers located in Kalimpong, Bijanbari and Mirik (Mukhopadhyay *et al.*, 1994). Mycorrhizal fungi colonize plant roots and extend the root system into the surrounding soil. Estimates of amounts of mycorrhizal filaments present in healthy soil are astounding. Several miles of filaments can be present in less than a thimbleful of soil associated with vigorously growing plants. The plant enjoys improved nutrient and water uptake, disease resistance and superior survival and growth.

Materials and Methods

Soil samples collected from the above sites were screened for fungal isolates using serial dilution and root maceration.

Isolation of Vesicular Arbuscular Mycorrhizal Spores : Spores of arbuscular mycorrhizal fungi were isolated from rhizosphere soil of mandarin by wet sieving and decanting method (Gerdemann and Nicolson, 1963). Approximately 250 ml of soil was suspended in 1 liter water. Heavier particles were allowed to settle for a few minutes and the liquid was decanted through sieves of decreasing size (BS 60, BS 80, BS 100, BS 150 and BS 200). Pores are fine enough to remove the larger particles of organic matter, but coarse enough to allow the desired spores to pass through. The suspension that passed through these sieve was saved and stirred to resuspend all particles. The heavier particles were allowed to settle for a few seconds and the liquid decanted again through the sieve and spores collected by fine brushes and were kept in different petriplates according to their size and colours. Further purification of AMF spores was done by sucrose gradient centrifugation method. In sucrose gradient centrifugation (Daniels and Skipper, 1982), spores and minimal amount of organic particles were further purified by suspending sieving in 40% sucrose solution and centrifuging at 2000 rpm (approximate 370 x g) for 1 minute. The supernatant (with spores) was passed through a sieve of BS 200 mesh and rinsed with distilled water to remove sucrose residue. With the help of a simple microscope (20X) parasitized spores, plant debris etc were separated and clean spores were stained with Poly Vinyl Lacto Glycerol and studied microscopically. For further use the spores were stored in Ringer's Solution (8.6gm NaCl, 0.3gm KCl, 0.33gm $CaCl\text{-}_2$ in 1 ltr. of boiled distilled water) at -15 to -20°C or in sterile distilled water.

Histopathology : Fungal association of AM fungi within the root tissues was observed according to Philips & Hayman (1970). Roots were cut into 1cm or smaller pieces and washed in tap water. It was boiled in 2% KOH in hot water bath for 1 hour. The KOH was decanted and the roots washed with water for 2-3 times. 1% HCL was added and kept for 30 minutes. After decanting the HCL the sample was washed thrice in tap water and cotton blue, lactic acid and glycerol was added in the ratio 1:1:1 to stain the internal structures of AMF inside the root segments i.e. arbuscules, vesicles, auxiliary cells, and boiled in water bath for 1 hour. The excess stain was decanted and sample placed in 50% glycerol for destaining. The roots were then crushed under pressure in slide and covered with cover slip for microscopic observation.

Serial Dilution :Fungal isolates were obtained by serial dilution method. Approximate 10 gram of soil was taken and suspended in 100 ml of sterile distilled water. It was agitated for an hour and then allowed to settle for another hour. The top aqueous liquid layer was taken and dilutions made upto 10^{-5}. 1 ml of each dilution was spread into petri plates containing potato dextrose agar (PDA) media. Replicates of 2 each was made for each dilution. The plates were observed everyday for formation of fungal colonies. As soon as fungal colonies appeared, they were isolated in PDA slants and preserved for identification and further use.

Evaluation of Phosphate Solubilizing Activity of Selected Isolates : Evaluation of phosphate solubilising activity of 10 isolates were done by growing the isolates in the two sets of Pikovs kaya's liquid medium amended with 0.5% tricalcium phosphate and 0.5% rock phosphate separately over a period of 10 days at 28°C with constant shaking at 100 rpm in a rotary incubator. Quantitative estimation of phosphate was done following ammonium molybdate ascorbic acid method as described by Kundsen and Beegle (1988). Amount of phosphate utilised or solubilised by the isolates were expressed as mg/L phosphate utilised by deducting the amount of residual total phosphate from the initial amount of phosphate source added to the modified Pikovskaya's liquid medium(yeast extract, 0.50 g/L, dextrose 10 g/L, calcium phosphate/rock phosphate 5.0 g/L, ammonium phosphate 0.50 g/L, potassium chloride 0.20 g/L, magnesium sulphate, 0.0001 g/L, ferrous sulphate 0.0001 g/L, pH 6.5)amended with 0.5% tricalcium phosphate and 0.5% rock phosphate. 50 ml of the liquid medium was inoculated with 5% v/v of the spore suspension prepared from the 7 days old culture grown on PDA slants and incubated at room temperature for 4 days with routine shaking at 100 rpm. The initial pH of the medium was recorded with a pH meter fitted with electrode. The mycelia were harvested after 10 days of incubation by filtering and the change in the pH of the culture filtrate was recorded after centrifuging the medium at 5000 x g for 5 mins. on a table centrifuge.

Preparation of Genomic DNA : Extraction of genomic DNA from isolates of mandarin were done. Fungal mycelia (3-4 day- old) was crushed with liquid nitrogen and incubated with lysis buffer containing 250 mM Tris-HCl (pH 8.0), 50 mM EDTA (pH8.0), 100 mM NaCl and 2% SDS, for 1 hr at 65°C followed by centrifugation at 12,000 rpm for 15 min. The supernatant was extracted with equal volume of water saturated phenol, centrifuged at 12,000 rpm for 15 min, and further extracted with equal volume of phenol : chloroform

: isoamyl alcohol (25:24:1) by centrifugation at 12000 rpm for 15 min; the aqueous phase was transferred in a fresh tube and chloroform (1:4 v/v) was added followed by 0.5M Na-acetate (1:10 v/v). Next Isopropanol was added to the above mixture (0.7 times the final volume) and centrifuged. DNA was precipitated from the aqueous phase with chilled ethanol (100%) and pelleted by centrifuging at 12000 rpm for 15 min followed by washing in 70% ethanol and centrifugation. The pellets were air dried and suspended in TE buffer pH 8.

Qualitative and quantitative estimation of DNA : Total genomic DNA extracted from isolates of mandarin was resuspended in 100 μl 1 X TE buffer, treated with RNAse (60 μg) and incubated at 37°C for 30 min After incubation the sample was re-extracted with PCI (Phenol: Chloroform: Isoamylalcohol 25:24:1) solution and RNA free DNA was precipitated with chilled ethanol as described earlier. The quality and quantity of DNA was analyzed both spectrophotometrically and in 0.8% agarose gel. The DNA from all isolates produced clear sharp bands, indicating good quality of DNA.

ITS-PCR amplification : All isolates of mandarin were taken up for ITS-PCR amplification. Genomic DNA was amplified by mixing the template DNA (50 ng), with the polymerase reaction buffer, dNTP mix, primers and Taq polymerase. Polymerase Chain Reaction was performed in a total volume of 100 ml, containing 78 ml deionized water, 10ml 10 X Taq pol buffer, 1 ml of 1 U Taq polymerase enzyme, 6 ml 2 mM dNTPs, 1.5 ml of 100 mM reverse and forward primers and 1 ml of 50 ng template DNA. The following primer pairs were used for ITS PCR (Table 1).

Table 1 : The nucleotide sequence used for ITS PCR

Seq Name	Primer Seq 5'-3'	Mer	TM	% GC
ITS 1	TCCGTAGGTGAACCTGCG	18	61	56%
ITS4	TCCTCCGCTTATTTGATATGC	21	63	59%
Fcg17F	TCGATATACCGTGCGATTTCC	21	65	47%
Fcg17R	TACAGACACCGTCAGGGGG	19	66	63%
T/ITS 1	TCTGTAGGTGAACCTGCGG	19	63.9	57%
T/ITS4	TCCTCCGCTTATTGATATGC	20	61.5	45%

PCR was programmed with an initial denaturing at 94°C for 5 min. followed by 30 cycles of denaturation at 94°C for 30 sec, annealing at 59°C for 30 sec and extension at 70°C for 2 min and the final extension at 72°C for 7 min in a Primus 96 advanced gradient Thermocycler. PCR product (20 μl) was mixed with loading buffer (8 μl) containing 0.25% bromophenol blue, 40 % w/v sucrose in water, and then loaded in 2% Agarose gel with 0.1 % ethidium bromide for examination with horizontal electrophoresis.

RAPD -PCR : For RAPD-PCR, the following nine random primers were selected (Table 2).

Table 2 : The nucleotide sequence used for RAPD

Seq Name	Primer Seq 5'-3'	Mer	TM	% GC
OPA-1	CAGGCCCTTC	10	38.2	70%
OPA-4	AATCGGGCTG	10	39.3	60%
OPD-2	TGATCCCTGG	10	34	60%
OPB-3	GATCCCCTG	10	37	70%
OPB-6	TGCTCTGCCC	10	40	70%
OPD-5	TGAGCGGACA	10	37	60%
A-5	AGGGGTCTTG	10	31.8	73%
AA-11	AGACGGCTCC	10	37.1	70%
AA-04	AGGACTGCTC	10	25.8	60%

PCR was programmed with an initial denaturing at 94°C for 4 min. followed by 35 cycles of denaturation at 94°C for 1 min, annealing at 36°C for 1 min and extension at 70°C for 90 s and the final extension at 72°C for 7 min in a Primus 96 advanced gradient Thermocycler. PCR product (20 μl) was mixed with loading buffer (8 μl) containing 0.25 % bromophenol blue, 40 % w/v sucrose in water, and then loaded in 2% Agarose gel with 0.1% ethidium bromide for examination by horizontal electrophoresis.

Scoring and data analysis

The image of the gel electrophoresis was documented through Bio-Profil Bio-1D gel documentation system and analysis software. All reproducible polymorphic bands were scored and analysed following UPGMA cluster analysis protocol and computed *In Silico* into similarity matrix using NTSYSpc (Numerical Taxonomy System Biostastiscs, version 2.11W) [Rohlf F.J. 1993]. The SIMQUAL program was used to calculate the Jaccard's coefficients. The RAPD patterns of each isolate was evaluated, assigning character state "1" to indicate the presence of band in the gel and "0" for its absence in the gel. Thus a data matrix was created which was used to calculate the Jaccard similarity coefficient for each pair wise comparison. Jaccard coefficients were clustered to generate dendograms using the SHAN clustering programme, selecting the unweighted pair-group methods with arithmetic average (UPGMA) algorithm in NTSYSpc.

Results and Discussion

Among the AM fungi, *Glomus mosseae* could be determined as the predominant one, followed by *Gigaspora*, *Acaulospora* and *Scutellospora*. Percentage of AM spores determined from different regions showed maximum of different *Glomus* sp., *Acaulospora* and *Scutellospora* in the foothills of Darjeeling, whereas maximum *Gigaspora* was obtained from the hilly areas (Table 3).

Glomus fasciculatum (Thaxt.) Gerd. & Trappe emend. C. Walker & Koske.

Colour pale yellow to bright brown with globose to subglobose in shape. Spores are produced directly with one or more subtending hyphae attached to it. Spore wall is continuous. Spore wall consisting of three layers (L1, L2, and L3). Spore size ranges from 70-120μm in diameter.

Glomus aggregatum N.C. Schenck & G.S. Sm. emend. Koske Spores globose to oval in shape. Size ranges from 40-120 μm in diameter, color-pale yellow. Formed singly or in sporocarps. Spore wall consist of 1-2 layers. Sporocarps are formed in loose clusters, from a single stalk, diameter ranges from 200-1800 x 200-1400 μm in size.

Glomus mosseae (Nicol. & Gerd.) Gerd. & Trappe

Brown to orange-brown in colour, shape, globose to sub-globose with an average diameter of 200 μm. Presence of three hyaline layers with subtending hyphae attached. Hyphae are double layered.

Glomus drummondii Blaszk. & C.

Renker Spores occur singly in the soil; develops from the tip of extraradical hyphae of mycorrhizal roots. Spores are golden yellow, globose to subglobose, average diameter is 70μm in diameter; single subtending hypha attached with the spore. Spore wall consists of three distinct layers.

Acaulospora bireticulata F.M. Rothwell & Trappe

Spores are globose and brownish in colour, diameter ranges from 280-410μm. Surface ornamentation is prominent. Spore wall consists of three layers. Spores are borne laterally from the neck of a sporiferous saccule.

Acaulospora capsicula Blaszk.

Spores are bright red or orange red in colour, globose to subglobose with three distinct layers. Layer 1, forming the spore surface, evanescent, hyaline, (0.8-) 1.5 (-2.5) μm thick, continuous with the wall of a sporiferous saccule, usually completely sloughed in mature spores. Layer 2 laminate, smooth, orange red to capsicum red 4.2μm thick. Layer 3 laminate, rigid, hyaline, 2.4 μm thick, easily separating from layer 2.

Gigaspora gigantea (Nicol. & Gerd.) Gerd. & Trappe Spores single in the soil; formed terminally or laterally on a bulbous sporogenous cell; greenish yellow (globose to subglobose; 300μm diam; sometimes ovoid; 250-270 x 265-370 μm. Subcellular structure of spores consists of a spore wall with two layers and one germinal wall.

Gigaspora margarita W.N. Becker & I.R. Hall

Spores produced singly in the soil, blastically at the tip of a bulbous sporogenous cell. Spores yellowish white to sunflower yellow; globose to subglobose; 357 μm diam; sometimes ovoid; 300-340 x 360-380 μm. Sporogenous cell orange to brownish yellow (clavate; 52.5-60.0 x 75.0-100.0 μm. Structure of sporogenous cell composed of two layers. Layer 1 hyaline, 1.7 μm thick approximately. Continuous with spore wall layer 1. Layer 2 orange to brownish yellow, 5.6 μm thick, continuous with spore wall layer 2.

Glomus clarum Nicol. & Smith

Spores single in the soil; hyaline to pale yellow, globose to subglobose; 150 μm diam; sometimes ovoid; 90-100 x 140-180 μm; with one subtending hypha, hyaline to pale yellow straight to curved;

wall of subtending hypha hyaline to pale yellow ,thick at the spore base; composed of three layers

Scutellospora pellucida (Nicol. & N.C. Schenck) C. Walker & F.E.Sanders

Spores single in the soil; formed terminally on a bulbous subtending hypha; hyaline to yolk yellow; globose to subglobose; 195μm diam; sometimes ovoid; 130-155 x 160-235 μm.

Scutellospora rubra (Nicol. & N.C. Schenck)

Spores color: dark orange-brown to red-brown, most tending toward the latter at maturity. Immature spores are white to cream with a rose tint under a dissecting microscope. Shape: globose to subglobose. Size 180 μm in average.

Besides, thirty five fungal isolates were obtained from the different soils of Darjeeling district. Among these isolates, four isolates showed formation of halo zone in PVK (Pikovakaya) medium indicating phosphate solubilising activity whereas all obtained isolates were estimated in liquid medium (Table 4). The morphological characteristic of isolated AMF and fungi are described below.(Fig -1)

Aspergillus niger

Colonies are black on PDA medium, mycelia are hyaline, aseptate, conidial heads radiate, conidia brown, ornamented with warts and ridges, subspherical, 3.5-5.0 mm diam. Conidiophore: consisting of a dense felt of conidiophores. Conidiophore stipes smooth-walled, hyaline. Vesicles subspherical, 50-100 mm diam

Aspergillus fumigatus

Colonies dark blue-green, consisting of a dense felt of conidiophores, intermingled with aerial hyphae. Mycelia hyaline, aseptate. Conidia verrucose, (sub) spherical, 2.5-3.0 μm dia Conidial heads columnar; conidiogenous cells uniseriate. Conidiophore stipes smooth-walled, often green in the upper part. Vesicles subclavate, 20-30 μm wide.

Aspergillus oryzae

Colonies growing rapidly, pale greenish-yellow, olive-yellow or with different shades of green, typically with dull brown shades with age. Mycelia hyaline, aseptate. Conidial heads radiate to loosely columnar, 150-300 μm diam. Conidia (sub)spherical to ovoidal, 4.5-8.0 (-10.0) x 4.5-7.0 μm, smooth-walled to roughened, greenish to brownish. Conidiophore stipes hyaline, up to 4-5 mm in length. Vesicles subspherical, up to 75 mm diam. Conidiogenous cells uniseriate and biseriate. Metulae or phialides covering the entire surface or the upper three-fourths of the vesicle.

Aspergillus versicolor

Colonies reaching 2-3 cm diam on CzA and 4-5 cm diam on MEA in two weeks at 25°C; variable in colour, light yellowish, pink to flesh-coloured, ochre or orange yellow to yellowish green, with exudate and reverse of equally variable colour. Mycelia hyaline, aseptate. Conidia globose, echinulate, mostly 2-3 μm diam. Conidiophores colourless or yellowish, smooth-walled, to 500-700 μm long; vesicles elongate with metulae and phialides covering most of the surface; conidial heads radiating.

Sporotrichum pruinosum

Distinct greyish or pinkish hue; Mycelia hyaline, aseptate. Blastoconidia from unbranched conidiophores ellipsoidal to ovoid pyriform or nearly cylindrical, 5.8 x 3.5 μm. All blastoconidia broadly attached and becoming thickwalled. Chlamydospores terminal or intercalary, hyaline, (sub) globose to broadly ellipsoidal or more rarely pyriform, 11- 60 μm diam or 11 x 7.5 μm, with granular contents and thick walls (up to 4.5 μm). Arthroconidia hyaline, cylindrical or rather irregular, often with granular contents, thin-walled, but sometimes becoming slightly thick-walled and more ellipsoidal. Conidiophores simple or typically branched. Branching racemose, each branch forming a terminal blastoconidium. Blastoconidia from branched conidiophores hyaline, subglobose to ellipsoidal or ovoidal, 10 x 8.5 μm.

Table 3 : Population of AM fungi in the rhizosphere of mandarin collected from different regions

Genus & species	% of AM spores from different regions			
	Kalimpong	Mirik	Bijanbari	Foothills
Glomus sp.				
G. aggregatum	44	23	48	82
G.mosseae	47	22	18	86
G. fasciculatum	45	30	38	82
G. viscosum	-	-	-	12
G. macrocarpum	05	08	08	10
G. microaggregatum	13	12	10	18
G. versiforme	-	-	-	06
Gigaspora sp.				
G decipiens	60	57	48	03
G. gigantea	75	78	82	14
G. margarita	65	72	70	34
G. rosea	25	20	-	08
Acaulospora sp.				
A. bireticulata	15	18	16	24
A. spinosa	08	12	10	22
A. denticulata	10	12	10	10
A. tuberculata	02	-	01	04
Scutellospora sp.				
S. rubra	06	08	06	12
S. calospora	04	02	04	05

Trichoderma harzianum

Dark green colony in PDA. Conidia. subglobose to ovoidal, 3.5 to 4.0 μm long, smooth, green, Conidiophores typically with paired branches forming over 150 μm of the length of terminal branches. Cells supporting the phialides equivalent in width to, or at most only slightly wider than, the base of phialides arising from them. Phialides: 6.5-6.7 μm long, 2.5-3.5 μm wide at the widest point 1.6-2.5 μm at the base; supporting cell 2.4-3.6 μm; Terminal phialides in a whorl or solitary, typically cylindrical or at least not conspicuously swollen in the middle and longer than the subterminal phialides.

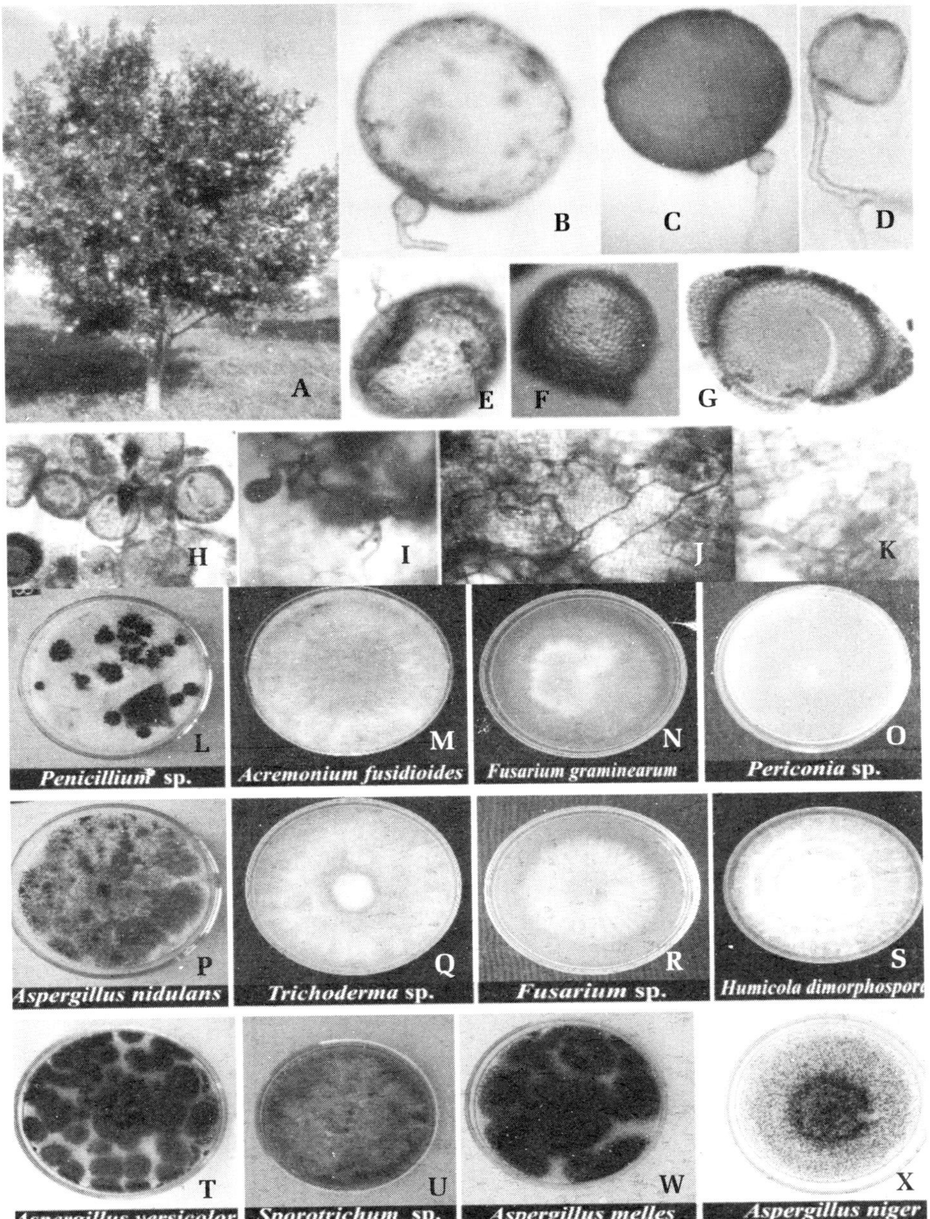

Figure 1 : *Citrus reticulata* (A), AMF spore of *Gigaspora decipiens* (B), *Scutellospora gregaria* (C), *Glomus mosseae* (D), *Gigaspora margarita* (E), *Acaulospora* sp. (F), *Entrophospora* sp. (G), *Glomus fasciculatum* (H), Histopathological studyof root tissues (I-K) and radial growth pattern of fungal isolates in PDA medium obtained from mandarin rhizophere (L-X).

Table 4 : Evaluation of phophphorus solubilization by fungal isolates obtained from forest, riverine and rhizosphere soil in liquid media amended with tricalcium phosphate (TCP) and rock phosphate (RP).

PSF	Isolates	TCP	RP
Aspergillus niger	FS/L04	850±2.58	366±3.76
	FS/L-40	852±4.41	370±322
	FS/G-140	824±3.06	344±3.25
	FS/C143	821±303	345±1.15
	FS/G-160	824±2.86	346±2.92
	FS/S-165	830±2.75	352±2.76
	FS/S-173	802±2.91	343±2.85
	FS/S-177	843±2.03	341±1.31
	FS/S-108	808±3.55	350±1.99
	FS/S-109	802±2.55	355±1.57
	FS/S110	842±2.92	367±1.38
	FS/S-112	842±2.76	354±1.09
	FS/S-113	848±2.85	360±0.92
	FS/S-262	795±1.31	360±1.30
Aspergillus melleus	FS/L-42	830±1.98	360±2.91
	FS/L-13	817±1.82	381±2.03
	FS/L-17	820±1.82	379±3.55±
	FS/L-18	821±3.07	376±2.55
	FS/G-226	847±1.70	352±2.92
	FS/S-64	842±1.86	211±2.76
	FS/L-41	843±2.36	214±2.85
	FS/S-63	839±1.54	332±1.31
	FS/S-24	810±2.51	338±1.99
	FS/S-278	829±3.17	339±1.57
Aspergillus clavatus	RS/P/14	852±1.99	360±1.38
	RS/D-288	830±1.57	350±1.09
	RS/T-57	809±1.38	352±0.92
	RS/T-58	802±1.09	354±1.30
Aspergillus nidulans	RS/T-59	830±2.86	350±2.92
	RS/P -60	840±2.36	340±2.76
Aspergillus fumigatus	RS/R-115	836±4.24	338±1.38
	RS/T-182	810±3.25	309±2.91
	RS/T-183	850±3.91	317±2.03
	RS/P-61	847±3.2	343±3.55
	RS/P05	854±3.25	370±2.55

PSF= Phosphate solubilizing fungi; TCP=Tricalcium phosphate (P=997 mg/L); RP= Rock phosphate (P=500 mg/L).Values are mean of 3 replicates.

Acremonium fusidioides

Colonies reaching 8-10 mm diam in ten days at 20°C on MEA, ochraceous-brown, powdery; hyaline and aseptate mucelia, conidia catenulate, of two kinds: (a) predominantly slightly pigmented, fusiform with truncate ends, 6.4 x 2.1 μm, (b) globose, hyaline, slightly warty, 3.4-4.7 μm diam.

Fusarium graminearum

Colonies reddish white, hyaline and aseptate mycelia, conidia often formed sparsely, falcate, sickle-shaped or markedly dorsi-ventral, 3-7-septate, 25-50 x 3-4 μm, with a well developed, often pedicellate foot cell.

Microconidia absent. Macroconidia produced from doliiform phialides 10-14 x 3,5-4,5 μm, formed laterally or on short multibranched conidiophores; sporodochia may form in older cultures. Chlamydospores, when present, are intercalary, single, in chains or clumps, globose, thick-walled, hyaline to pale brown with a smooth or slightly roughened outer wall, 10-12 diam. Many strains fail to develop chlamydospores on standard media.

Humicola dimorphospora

Aseptate and hyaline mycelia, sympodial conidia mostly guttuliform, 3-8 (4.3 $\pm$ 0.8) x 1.5-3 (2.2 $\pm$ 0.3) μm, hyaline to subhyaline, smooth- and thin-walled. Lateral blastoconidia from undifferentiated hyphae mostly subglobose to obovoid, 3-5(4 $\pm$ 0.4) x 3-5(3.8 $\pm$ 0.4) μm, smooth- and thick-walled, brown, always becoming dark gradually, sessile or on subhyaline to weakly pigmented pedicels. Conidiophores more or less differentiated, mostly unbranched, up to 90 μm long. Conidiogenous cells terminal or intercalary, sympodial, denticulate, 22-43 long, 1-1.5 μm wide toward the center, often inflated at the apex up to 5 μm wide.

Amplification of target DNA through PCR with sequence specific primers is potentially more sensitive and rapid than microbiological techniques, as a number of constrains are removed. Unlike culture, PCR does not require the presence of viable organisms for success and may be performed even when sample volumes are small.

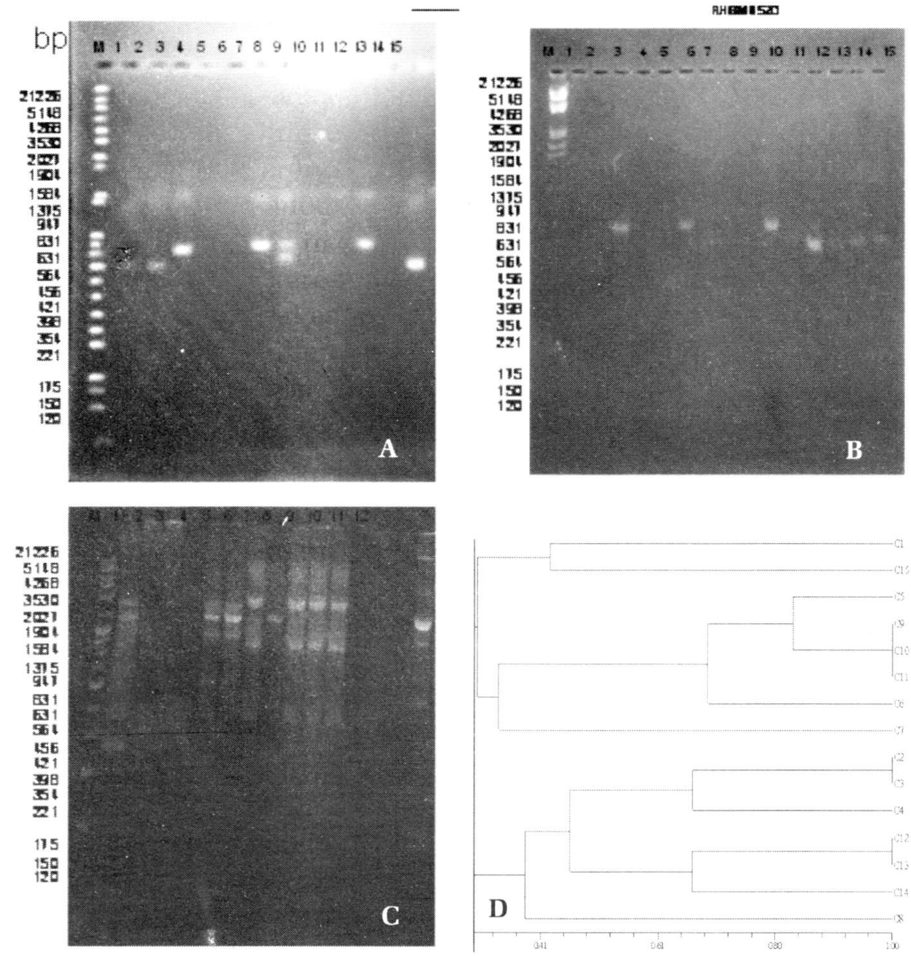

Figure 2 : A&B. Agarose gel (2%) electrophoresis of ITS region. Agarose gel (2%) electrophoresis of RAPD. **C**.Lane M-high range DNA ladder. D. The phylogenetic analyses conducted using the UPGMA method by NTSYS software.

Differences in the nucleotide composition of the variable ITS region have been successfully employed to design specific primer sets that amplify DNA selectively among and within species of plant pathogens (Nazar *et al*., 1991; Moukhamedove *et al*., 1994; Schilling *et al*., 1996; Moricca *et al*., 1998). In the broader context, taxon-selective amplification of ITS regions is likely to become a common approach in molecular identification strategies, ITS regions have been used successfully to generate specific primers capable of differentiating closely related fungal species (Bryan *et al.*, 1995). These rDNA are highly stable and exhibit a mosaic of conserved and diverse regions within the genome (Hibbett, 1992). They also occur in multiple copies with up to 200 copies per haploid genome (Bruns et al.1991) arranged in tandem repeats with each repeat consisting of the 18S small subunit (SSU), the 5.8S, and the 28S large subunit (LSU) genes.

Fungal isolates of mandarin rhizosphere produced a single amplified product ranging from 600 bp to 800 bp (Fig. 2 A-B) which is in accordance with Mukherjee *et al.* (2002) who studied the identification and genetic variability of the mandarin isolates. These results are also in accordance with several workers who observed the amplified rDNA fragment of approximately 600 bp by ITS-PCR in mandarin (Ospina *et al.*, 1999; Muthumeenakshi *et al.*, 1994; Lieckfiledt *et al.*, 1999; Venkateswarlu *et al.*, 2008). The genetic relatedness among isolates of mandarin, were analyzed by random primers OPA-1; OPA-4; OPB2, OPB3, OPD-6; OPD5, A-5; AA-04 and AA-11 to generate reproducible polymorphisms. All amplified products with the primers had shown polymorphic and distinguishable banding patterns which indicate the genetic diversity of all isolates. A total of 73 reproducible and scorable polymorphic bands ranging from approximately 100bp to 2000bp were generated with primers among the nineteen mandarin isolates . In the RAPD profiles showed that primer A-5 scored highest bands which ranged between 200bp to 6000bp in all cases (Fig. 2 C). Relationships among the isolates were evaluated by cluster analysis of the data based on the similarity matrix. The Dendogram (Fig. 2 D) was generated by unweighted pair-group methods with arithmetic mean (UPGMA) using NTSYSpc software.

References

Bruns, T.D., White, T.J., and Talyor, J.W. (1991). Fungal molecular systematics. *Annu. Rev. Ecol. Sys*, 22: 525-564.

Daniels, B. A. and Skipper,H. D. (1982). Methods for the recovery and quantitative estimation of propagules from soil. p. 29-35. In: N. Schenck(eds.), Methods and Principles of mycorrhizal Research Amer. Phytopathol. Soc., St. Paul, Minnesota.

Gerdemann, J. W. and Nicolson, T. H. (1963). Spores of mycorrhizal *Endogone* species extracted from soil by wet sieving and decanting. *Trans. Br. Mycol. Soc.* 46: 235-244.

Hibbett, D.S. (1992). Ribosomal RNA and fungal systematics. *Trans. Mycol. Soc.* 33: 533-556.

Knudsen, D. Beegle, D. (1988). Recommended phosphorous tests.122-125.In: W.C. Dahnke (ed), Recommended chemical soil tests procedures for the north central region. Bull North Dakota Agric Exp Stn. North Dakota, USA No. 499.

Lieckfiledt, E., Samules G.J., Helgard I., Nirenderg and Petrini O. (1999). A morphological and molecular perspectives of *T.viride* is it one or two species? *Appl Environ Microbiol,* 65(6):2418-2428.

Moukahmedov, R., Hu, X., Nazar, R.N. and Robb, J., (1994). Use of polymerase chain reaction-amplified ribosomal intergenic sequence for diagnosis of *Verticillium tricorpus*. *Phytopathology* 84: 256-259.

Mukherjee, P.K, Verma, A. and Latha, J., (2002). PCR fingerprinting of some *Trichoderma* isolates fom two Indian type culture collection- a need for re-identification of these economically importance fungi. *Science Correspondence,* 83(4): 372-374.

Mukhopadhyay P, (1994). The state of mandarin orange cultivation in Darjeeling district. *Rev. Trop. Pl. Path.* 2(1985/86) : 95-105.

Rohlf, F.J. (1993). NTSYSpc: Numerical Taxonomy and Multivariate analysis system. Ver 2.11 W. Exteer Software: Setauket, N Y. *Science and Culture* 67(1-2): 42-46.

Moricca, S., Ragazzi, A., Kasuga, T., Mitchelson, K.R., (1998). Detection of *Fusarium oxysporum* f. sp. *vasinfectum* in cotton tissue by polymerase chain reaction. *Plant Pathol.* 47: 486-494.

Venkateswarlu, R., Reddi, K.M., Reddy, E.N.P, and Sudhakar P. (2008). Molecular characterization of *Trichoderma* spp. Used against *Fusarium* wilt with PCR based RAPD and ITS-PCR. *Journal of Mycology and Plant Pathology.* 38(3): 569-563.

□□□

Chapter **13**

MOLECULAR DYNAMICS SIMULATION OF PR-1 PROTEIN FROM *SOLANUM TUBEROSUM* PROVIDES AN IN-DEPTH VIEW OF ITS STRUCTURAL FEATURES

SUBARNA THAKUR[1], ASIM K BOTHRA[2], SAUBASHYA SUR[1] AND ARNAB SEN[1]

[1] NBU Bioinformatics Facility, Department of Botany, University of North Bengal, Siliguri 734013,India

[2] Bioinformatics Chemoinformatics Laboratory, Department of Chemistry, Raiganj University College, Raiganj, India

Introduction

The pathogenesis-related proteins are a family of plant proteins encoded by the host plant but induced only in major pathological or related situations. PR proteins can be classified into 17 major groups on the basis of their sequence characteristics, immunological relationships and biological activities (Liu, 2005). Among the various PR proteins, PR-1 is the most abundant and reaching upto 1-2% of the total leaf proteins. Proteins belonging to PR-1 family were first identified from tobacco-mosaic virus infected tobacco plants (Van loon, 1976). Since then, many proteins with similar properties have been isolated from many other dicot and monocot plant species. Although these proteins have been shown to have limited antifungal activities, suggesting a potential role in plant defense, their molecular mechanism and mode of action are still relatively unknown. The physicochemical properties of these proteins enable them to survive in the harsh environmental conditions where they are found such as vacuolar compartment or intercellular spaces (Niderman *et al.*, 1995). Although structures of some pathogenesis related proteins associated with tomato (Ghosh and Chakrabarti, 2005), (Fernandez *et al.*, 1997) and tobacco (Koiwa *et al.*, 1999) have been reportedly resolved by X-ray crystallography and NMR techniques, almost no such work has been reported on the 3-D structure of pathogenesis related proteins particularly PR-1 of *S. tuberosum* (potato).

In the present study, we have constructed the three dimensional model of PR-1 protein based on the NMR structure of pathogenesis related protein P14a (PDB code-1CFE) by homology modeling method. A suitable refined model of the protein was then subsequently used for molecular dynamics simulation to understand the motional and structural properties which can then be interpreted to understand the functionality and biological activity of these proteins.

Materials and Methods

Homology modeling : The amino acid sequence of the PR-1 protein of *S. tuberosum* bearing accession number CAB58263 was obtained from the NCBI (National Center for Biotechnology Information) database (www.ncbi.nlm.nih.gov). The first step of model building was to search for a number of related sequences to find a related protein that will be used as a template. Position specific iterative blast (PSI-BLAST) (Altschul *et al.*, 1997) (http://www.ncbi.nlm.nih.gov/blast/) was carried out against database specification of PDB proteins to detect similarity. Template selection was based upon its environmental and functional resemblance as well as phylogenetic similarity with the protein under consideration. The program ClustalW 1.83 (Thompson *et al.*, 1994) with default settings was used to carry out optimal alignment between the target sequence and template. Based on this alignment crude 3D model of the protein was created using the MODELLER9v4 program (Sali and Blundell, 1993). The crude model obtained is often erroneous and therefore refinement is indispensable. The constructed model was energetically minimized using the steepest descent (SD) and conjugate gradient (CG) methods. The computations were carried out in vacuo with the GROMOS96 43B1 factors using the Swiss-Pdb Viewer package (http://expasy.org/spdv/program/spdv37sp5.zip) (Kaplan and Littlejohn, 2001). Hydrogen bonds were ignored. The secondary structure of the protein was predicted with the help of HNN (Hierarchical Neural Network) methods.

In order to test the internal quality and reliability, the model was subjected to some evaluations: ProSA (Wiederstein and Sippl, 2007) was performed to judge the accuracy and of the modelled structures and check the 3D models for potential errors. VERIFY3D (Eisenberg *et al.*, 1997) was used to authenticate the refined structures. The Auto Deposition Input Tool (ADIT) (http://deposit.pdb.org/validate) was used to construct Ramachandran Plot to look into the possible conformations of θ and Ψ angles for our modeled protein. (Ramachandran *et al.*, 1963). SAVES (Structure analysis and verification server) (http://nihserver.mbi.ucla.edu/SAVS/) was used to verify the model quality. Presence of ligand binding pockets in the structures was predicted using CASTp server (Dundas *et al.*, 2006).

Molecular dynamics simulation : The constructed refined model of the PR protein of *S. tuberosum* was used as starting structure for the Molecular dynamics simulation. The simulation was performed using the Groningen Machine for Chemical Simulation (GROMACS) program (Lindahl *et al.*, 2001) running on linux platform. The starting structure was immersed in a cubic box with dimension of 10 Å consisting of SPC water molecules. The GROMOS96 force field parameters were used for the simulation. The Linear Constraint algorithm was used to constrain bond lengths, with a time step of 2 fs (Hess *et al.*, 1997). The simulation was conducted at constant temperature (300 K), coupling each component separately to a temperature bath using the Berendsen coupling method (Berendsen *et al.*, 1984). Nonbonded cutoffs of 1.0 nm were used for both Lennard-Jones and Coulomb potentials. The total MD simulation was performed for twelve nano seconds. Before running the simulation, the solvent was relaxed by energy minimization where 1000 steps of steepest descent (converging in 128 steps) was used followed by 1000 steps of conjugant gradient process (converging in 10 steps) and this was followed by 3.0 ns of simulation imposing positional restraints on the non-H atoms. The positional restraints were then released and 12 ns production run were obtained and analyzed. The trajectory files obtained during the whole simulation run was analyzed and RMSD, RMSF, gyration radius, energy potential were calculated with the help of GROMACS package. The time evolution of secondary structure was evaluated with help DSSP package (Kabsch and Sander, 1983).

Results and Discussion

Modeling : The 3D model of the protein was built taking PDB structure 1CFE (Fernández *et al.*, 1997) as the template protein. The template is a 135 amino acid residue PR protein, P14a from *Lycopersicon esculentum*. The P14a protein shows a 90% identity with PR-1 protein from *Solanum tuberosum*. The pair wise alignment reveals a high degree of conservation in the amino-acid residues amongst the two proteins. The major conserved regions lay between residues 25-51, 53-64, 66-82, 84-96 and 117-154. In the PR protein of *L. esculentum* (PDB code 1CFE) the residues Ser 3, Gln 5, His 48, Ser 49, His 903, Arg 100 and Asn 114 are strictly conserved regions of which His 48, Ser 49 and His 93 are functionally important sites. In our modeled protein the functionally important residues i.e., active sites corresponded to histidine in 72 position, serine in 73 position and histidine in 117 position.

Figure 1 shows the 3D structure of the modeled protein. The modeled protein was named as 1SOP. The 3D structure of the protein shows a novel α-β-α arrangement of α-helices and β-strands quite similar to that of 1CFE. The helix and sheets remain spread all through the protein structure. The α-β-α sandwich molecular architecture of the protein represents a rigid scaffold offering higher stability.

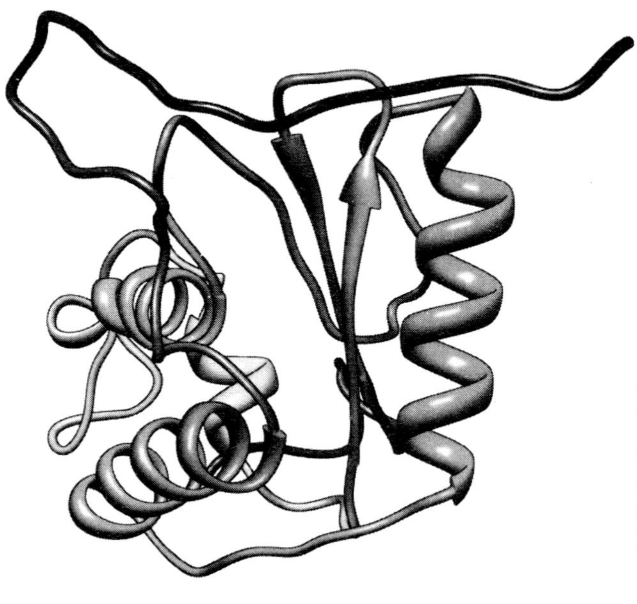

Figure 1 : Three dimensional structure of PR-1 of *Solanum tuberosum*. The structure has four alpha helices and beta sheet interspersed between them.

The higher stability is pivotal for the working of the PR-1 in extreme extra cellular environment. The structure HNN analysis of the secondary structure of the protein illustrated that the alpha helix portion consisted of 39 (24.53%), extended strand 28 (17.61%) and random coil 92 (57.86%) residues. The modeled structure consists of 1216 atoms and 1250 bonds. It had a molecular formula of $C_{754}H_{1136}N_{226}O_{225}S_{11}$.

The quality of the refined model assessed by different programs like ProSA, Verify 3D and ERRAT Score suggests that the predicted protein model is quite reliable and consistent. CASTp revealed the presence of 16 pockets for binding regions with varying area and volume. These pockets play an important role in the protein functionality. The Ramachandran plot representing the backbone conformations for the modeled protein reveals that 82.7 % of the non-glycine and non-proline residues of the modeled protein reside in the most favored regions.

Dynamics study : Several parameters such as RMSD, Radius of gyration (Rg), and RMSF etc were considered to assess the structural stability of the protein during the course of simulation. The RMSD was used to evaluate the deviation of the structure from the original starting structure over the period of the simulation. The figure 2 shows a plot of RMSD over time. It is quite apparent from the figure that RMSD gradually increases upto 3000 ps and then suddenly decreases around 4000 ps then again start increasing. From 7000 ps onwards it is observed that there is a leveling off in deviation from the reference structure. The time evolution of Radius of gyration (Rg) gives us an idea about the compactness of the structure. There is quite a variation in the Radius of gyration during the simulation particularly during 3000 to 5000 ps time frame. This further proves that a conformational change is taking place during the simulation period of 3000 to 5000 ps.

The Root mean square fluctuation is a measure of flexibility of each residue over the trajectory. The figure 3 demonstrates the RMSF of $C\alpha$ atoms plotted against the residue number over the time period 7000-12000 ps. From the plot it is obvious that the first few residues of the proteins fluctuate considerably. Importantly a major fluctuation is noticed in the residue stretches 5-20, 110-120 and 130-140 during the simulation reflecting that this region of the protein shows most mobility. Apart from the first few residues, a profound fluctuation is observed in residues 110-120. Some of these residues are part of the alpha helix IV secondary structure of the protein. During the simulation the alpha helices particularly the fourth helix shows major conformational changes. The flexibility of these residues can be attributed to their role in the functionality of the protein perhaps they constitute the binding sites of the protein. It also reemphasizes the fact that alpha IV helix plays a vital role in the α-β-α sandwich arrangement. The secondary structure analysis was done using the DSSP program. The simulation time frame of 7000 ps to 12000 ps was selected to examine the changes in the secondary structure because the protein is structurally quite stable in this period as revealed by RMSD and Radius of gyration values. The figure reveals that three alpha helices are relatively quite stable during the simulation and don't undergo any structural changes but the alpha helix comprising of the residues 115-120 residues is significantly broken during this period and is never recovered. The ß sheet present in the structure is also very much stable during the simulation. During the time course of the simulation, the total potential energy of the protein did not exceed -9.6×10^5 KJ mol^{-1}, indicating that the protein is energetically quite stable.

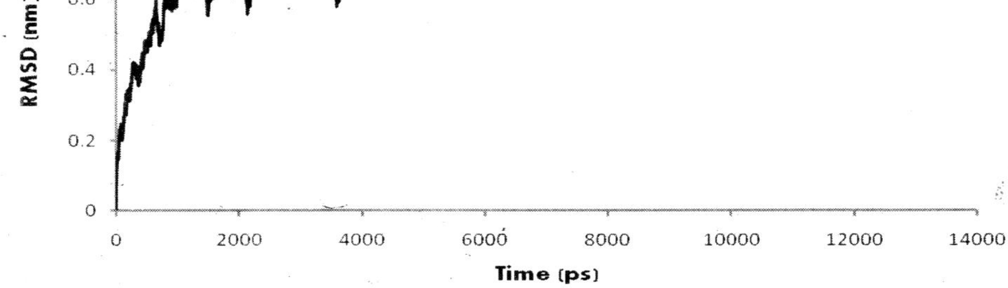

Figure 2 : The time evolution of RMSD during the whole simulation.

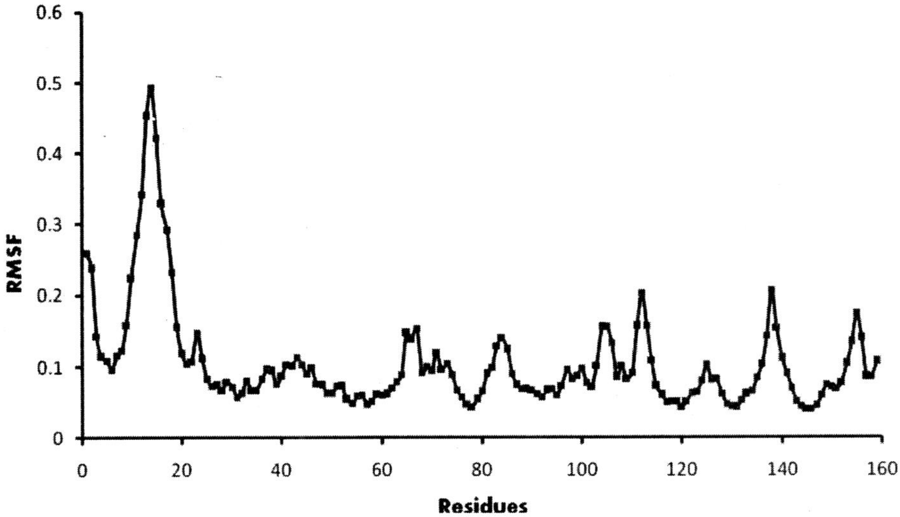

Figure 3 : RMS fluctuation in the Cα atoms of the residues of the protein in aqueous medium.

Conclusion

The three dimensional structure of the PR-1 protein from *S. tuberosum* provided an insight into its conformational properties and structure-function relationship. The structure presented here is quite consistent with their biochemical and motional properties. Molecular dynamics analysis revealed higher degree of stability but the residues of alpha IV helix are quite flexible during the simulation. The flexibility together with conservation of the residues 115-120 suggests their role as functionally important site. These results are vital for describing the mode of action of PR1 of plants and can be further substantiated with experimental evidences.

References

Altschul SF, Madden TL, Schaffer AA, Zhang J, Zhang Z, Miller W, and Lipman DJ. (1997). Gapped BLAST and PSIBLAST: a new generation of protein database search programs. *Nucleic Acids Res.*, 25(17):3389-3402.

Berendsen HJC, Postma JPM, van Gunsteren WF, DiNola A and Haak JR. (1984). Molecular dynamics with coupling to an external bath. *J. Chem. Phys.*, 81(8):3684–3690.

Dundas J, Ouyang Z, Tseng J, Binkowski A, Turpaz Y, and Liang J. (2006). CASTp: computed atlas of surface topography of proteins with structural and topographical mapping of functionally annotated residues. *Nucleic Acids Res.*, 34(Web Server issue):W116-W118.

Eisenberg D, Luthy R, and Bowie (JU) (1997). VERIFY3D: Assessment of protein models with three-dimensional profiles. *Metho. Enzymol.*, 277:396-404.

Fernandez C, Szyperski T, Bruyere T, Ramage P, Mosinger E, and Wuthrich K. (1997). NMR Solution Structure of the Pathogenesis-related Protein P14a. *J. Mol. Biol.*, 266(3):576-593.

Ghosh R and Chakrabarti C. (2005). Crystallization and preliminary X-ray diffraction studies of NP24-I, an isoform of a thaumatin-like protein from ripe tomato fruits. *Acta Crystallogr., Sect. F: Struct. Biol. Cryst. Commun.*, 61(8): 806-807.

Hess B, Bekker H, Berendsen HJC and Fraaije JGEM. (1997). LINCS: a linear constraint solver for molecular simulations. *J. Comput. Chem.*, 18(12):1463–1472.

Kabsch W, Sander C. (1983). Dictionary of protein secondary structure: pattern recognition of hydrogen-bonded and geometrical features. *Biopolymers.*, 22(12):2577-2637.

Kaplan W and Littlejohn TG. (2001). Swiss-PDB Viewer (Deep View). *Brief Bioinform.*, 2(2): 195-197.

Koiwa H, Kato H, Nakatsu T, Oda J, Yamada Y and Sato F. (1999).Crystal structure of tobacco PR-5d protein at 1.8 Å resolution reveals a conserved acidic cleft structure in antifungal thaumatin-like proteins. *J. Mol. Biol.*, 286(4):1137-1145.

Lindahl E, Hess B, Vander and Spoel D. (2001). GROMACS 3.0: A package for molecular simulation and trajectory analysis. *J.Mol. Model.*, 7(8):306-307.

Liu Z, Du L, Wan B. (2005). Pathogenesis-related proteins in higher plants. *Natur. Prod. Res. Dev.*, 17(1): 229–234.

Niderman T, Genetet I, Bruyère T, Gees R, Stintzi A, Legrand M, Fritig B, and Mosinger E. (1995). lsolation and Characterization of Three 14-Kilodalton Proteins of Tomato and of a Basic PR-1 of Tobacco with lnhibitory Activity against *Phyfophfhora infestans. Plant Physiol.* 108(2): 17-27.

RamachandranGN,Ramakrishnan C, and Sasisekharan, V. (1963). Stereochemistry of polypeptide chain configurations. *J Mol.Biol.*, 7(1):95-99.

Sali A, and Blundell TL. (1993). Comparative protein modelling by satisfaction of spatial restraints, *J. Mol.Biol.*, 234(3):283-291.

Thompson JD, Higgins DG, and Gibson T. (1994). CLUSTAL W: improving the sensitivity of progressive multiple sequence alignment through sequence weighting, position-specific gap penalties and weight matrix choice. *Nucleic Acids Res.*, 22(22):4673-4680.

Van Loon LC. (1976). Specific soluble leaf proteins in virus-infected tobacco plants are not normal constituents. *J Gen. Virol.* 30(3): 375–379.

Van Loon, L. C. and E. A. and Van Strein (1999). The families of pathogenesis-related proteins, their activities, and comparative analysis of PR-1 type proteins. *Physiol. Mol. Plant Pathol.*, 55(2) :85-97.

Wiederstein M, and Sippl MJ. (2007). ProSA-web: interactive web service for the recognition of errors in three-dimensional structures of proteins. Nucleic Acids Res., 35: W407- W410.

□□□

Chapter 14

ABIOTIC STRESS INDUCED BIOCHEMICAL RESPONSE IN CEREALS AND LEGUMES AND ASSOCIATED CHANGES IN ARBUSCULAR MYCORRHIZAL POPULATION

U. CHAKRABORTY[1], D. PRADHAN[1], R. LAMA[1], B. PRADHAN[1], U. DE[2] AND B.N.CHAKRABORTY[2]

[1] Plant Biochemistry Laboratory ; [2] Immuno-Phytopathology Laboratory, Department of Botany, University of North Bengal, Siliguri, West Bengal.

Introduction

Legumes and cereals occupy an important position in human diet as they are the major source of protein and nutrient. Abiotic stresses, such as temperature, and water deficiency are major factors which reduce crop productivity. Temperature stress is one of the major limiting factors in the production of legumes like soybean [*Glycine max* (L.) Merr.] and lentils (*Lens culinaris* Medikus); similarly, water stress is another major limiting factor for cereals like maize (*Zea mays* L.) and wheat (*Triticum aestivum L.)* .Temperature stress as heat, cold or freezing is a principle cause for yield reduction in crops (Boyer, 1982) and ROS (reactive oxygen species) generated by these stresses have been shown to injure cell membranes and proteins which lead to oxidative stress (Larkindale and Kright, 2002). Drought is a meteorological term and is commonly defined as a period without significant rainfall. Generally drought stress occurs when the available water in the soil is reduced and atmospheric conditions cause continuous loss of water by transpiration or evaporation. Drought stress tolerance is seen in almost all plants but its extent varies from species to species and even within species. Arbuscular mycorrhizal

(AM) fungi are known to improve the availability of nutrients especially phosphorus, moving them to the root surface by diffusion (Jakobsen *et al.*, 1992). It is also reported that mycorrhizal symbiosis enhances absorption of relatively immobile micronutrients such as Zn and Cu (Faber *et al.*, 1990; Kothari *et al.*, 1991; Li *et al.*, 1991; Liu *et al.*, 2000; Ryan and Angus, 2003). Increased population of AM fungi during stress increases the ability of plants to withstand the stresses.

Keeping these in mind, the present study was undertaken to investigate how legumes and cereals respond to temperature and water stress respectively with special reference to their antioxidative mechanisms and change in arbuscular mycorrhizal population during both stresses.

Materials and Methods

Seeds of soybean (cultivar JS 71-05) was collected form the National Centre of Soybean Research , Indore, M.P and Seeds of lentil (cultivar Asha) was collected from Pulses and Oil Research Centre, Berhampore. The seeds of maize and wheat were obtained locally.Seeds were soaked overnight in distilled water after surface sterilization with 0.1% $HgCl_2$ and grown in petriplates. For experimental purposes small seedlings were transferred to pots containing sandy loam soil mixed with farmyard manure. Plants were watered regularly and maintained properly. One month old seedlings of legumes and cereals were exposed to temperature and water stress respectively.

Antioxidant enzyme extraction and assays : For extraction of enzymes, leaf samples were initially ground to powder in liquid nitrogen and then extracted with 0.05 M sodium phosphate buffer, pH 6.8-7.2 (peroxidase, ascorbate peroxidase and catalase) and 0.1 M potassium phosphate buffer, pH 7.6 (glutathione reductase and superoxide dismutase) using polyvinylpyrrolidone under ice cold conditions. The homogenates were then centrifuged at 10,000 rpm for 15 min. Supernatants were used as crude enzyme extracts.

Assay

Peroxidase (POX: EC. 1.11.17) : For determination of peroxidase activity, 100μl of freshly prepared crude enzyme extract was added to the reaction mixture containing 1ml of 0.2M Sodium phosphate buffer (pH 5.4), 45 μl of 30% H_2O_2, 100 μl of O-dianisidine (5 mg/ml methanol) and 1.7ml of distilled water. Peroxidase activity was assayed spectrophotometrically in UV VIS spectrophotometer (Model 118 SYSTRONICS) at 460 nm by following the method of Chakraborty *et al.* (1993).

Ascorbate peroxidase (APOX: EC.1.11.1.11) : Activity was assayed as decrease in absorbance by monitoring the oxidation of ascorbate at 290 nm according to the method of Asada *et al.* (1987) with some modification. The reaction mixture consisted of 0.01 ml each of enzyme extract, 0.5mM ascorbic acid, H_2O_2 and 2.97ml of sodium phosphate buffer (pH 7.2).

Catalase (CAT: EC.1.11.1.6) : Catalase activity was assayed as described by Chance and Machly (1955). Enzyme extract (0.1 ml) was added to 3ml of H_2O_2 in phosphate buffer (0.4ml of H_2O_2 to 100 ml of phosphate buffer, pH 7.0) and the breakdown of H_2O_2 was measured at 240 nm in a spectrophotometer.

Superoxide dismutase (SOD: EC 1.15.1.1) :Activity was assayed by monitoring the inhibition of the photochemical reduction of nitroblue tetrazolium chloride (NBT) according to the method of Dhindsa *et al.*(1981) with some modification. Each 3ml of the assay mixture constituted of 0.1ml enzyme extract, 1.5ml phosphate buffer (0.1M, pH 7.6), 0.1ml Na_2CO_3 (1.5M), 0.1ml NBT (2.25mM), 0.2ml ml methionine (0.2M), 0.1ml EDTA (3 mM), 0.1 ml riboflavin (60 μM) and 0.8 ml of distilled water. The reaction tubes containing enzyme samples were illuminated with 15W fluorescent lamp for 10min. The other set of tubes lacking enzymes were also illuminated and served as control. A non-irradiated complete reaction mixture served as blank. The absorbance of samples were measured at 560 nm and 1 unit and activity was defined as the amount of enzyme required to inhibit 50% of the NBT reduction rate in the controls containing no enzymes.

Glutathione reductase (GR: EC 1.6.4.2) : Glutathione reductase activity was determined by the oxidation of NADPH at 340 nm as described by Lee and Lee (2000). The reaction mixture consisted of 0.1M potassium phosphate buffer (pH 7.6), 0.1M EDTA, 0.1mM NADPH, 0.6mM glutathione (oxidised form, GSSG) with an appropriate volume of enzyme extract. The reaction was initiated by addition of NADPH at 25^{0}C.

Extraction and estimation of Protein : Soluble proteins were extracted from leaves and roots by following the method of Chakraborty *et al.* (1995) and estimated following the method of Lowry *et al.* (1951), using BSA as standard.

Extraction of Phenolics : Phenol was extracted by following the method of Mahadevan and Sridhar (1982). One gram of leaf tissue was immersed in boiling absolute alcohol at the rate of 5ml alcohol per gram of tissue for 5-10 min, cooled, crushed in mortar and pestle using 80% alcohol in dark and filtered in dark chamber. The residues are re-extracted with 80% alcohol, and then the final volume was made up with 80% alcohol to 10ml.

Arnow's method (1933) was followed to estimate orthodihydroxy phenol in which 2ml of 0.5(N) HCl, 1ml of Arnow's reagent (10g $NaNO_2$ + 10g Na_2MO_4 in 100ml of distilled water) and 2ml of 1(N) NaOH were added in 1ml of crude extract. After reaction pink colour was developed and the volume was raised up to 10ml with distilled water and after shaking the absorbance was measured in colorimeter at 515nm.

Total phenol was estimated by following the method of Bray and Thorpe (1954). In one ml of extract, 1ml of 50% diluted Folin ciocalteu's and 2ml of 20% Na_2CO_3 solution were added. Then the mixture was boiled in water bath for 1minute. After cooling under running tap water the mixture was diluted with distilled water and final volume was made 25ml. The absorbance was measured at 650nm in colorimeter.

Extraction and estimation of Proline : Proline was extracted and estimated by following the method of Bates *et al.* (1973). 0.5g of plant tissue was crushed with 5ml of 3% Sulphosalicylic acid in mortar and pestle then was filtered through whatman No.1 filter paper at room temperature. The supernatant was collected for estimation.

For estimation to 1ml of extract, 3ml of distilled water and 1ml of Ninhydrin solution were added. The mixture was kept on a boiling water bath for 30 minutes. After cooling, the reaction mixture was transferred in a separating funnel and 5ml of Toluene was added and mixed vigorously. Lower coloured solution was taken and O.D. values were measured at 520 nm and quantify from a standard curve of proline.

Isolation of Arbuscular Mycorrhizae : Spores of arbuscular mycorrhizal fungi were isolated from rhizosphere soil by wet sieving and decanting method (Gerdemann and Nicolson, 1963). Approximately 250 gm of soil was suspended in 1 litre water. Heavier particles were allowed to settle for a few seconds and the liquid was decanted through sieves of decreasing size (BS 60, BS 80, BS 100, BS 150 and BS 200). Pores were fine enough to remove the larger particles of organic matter, but coarse enough to allow the desired spores to pass through. The suspension that passed through these sieves was saved and stirred to resuspend all particles. The heavier particles were allowed to settle for a few seconds and the liquid decanted again through the sieve and spores collected by fine brushes and were kept in different petriplates according to their size and colours. Moreover for further observations or purification of AMF spores sucrose gradient centrifugation method was used. In sucrose gradient centrifugation (Daniels and Skipper, 1982), spores and minimal amount of organic particles were further purified by suspending sieving in 40% sucrose solution and centrifuging at 2000 rpm (approximate 370 x g) for 1 minute. The supernatant (with spores) was passed through a sieve of 400 meshes and rinsed with distilled water to remove sucrose residue. With the help of a simple microscope (20X) parasitized spores, plant debris etc. were separated and clean spores were stained with Melzar's reagent and studied microscopically. For further use the spores were stored in Ringer's Solution at -15^{0}C.

Results

The study showed that during temperature stress the protein content showed a decrease in case of soybean during cold stress at 5°C when compared to the control while during the heat stress of 30°C in lentil, the protein content showed an increase. In maize water stress led to an increase in protein content, but in wheat accumulation decreased (Fig.1).

Proline content showed an increase in the leaves of both the legumes under elevated and low temperature stresses. Proline content of maize leaves also increased following water stress during the 5th day but wheat leaves did not show any significant change. The content

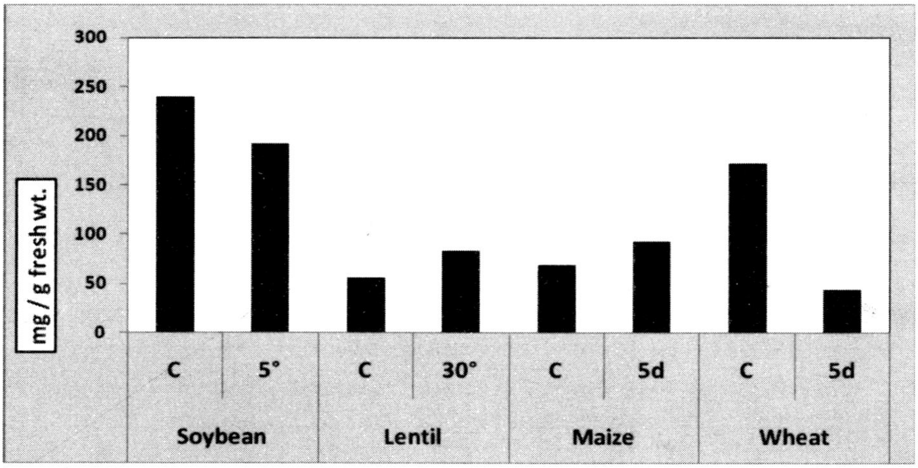

Figure 1 : Protein content: Soybean (cold stress) -control (25°C), Treated (5°C) Lentil (heat stress) -control (19°C), Treated (30°C) Maize & Wheat (water stress) - control, 5 day drought.

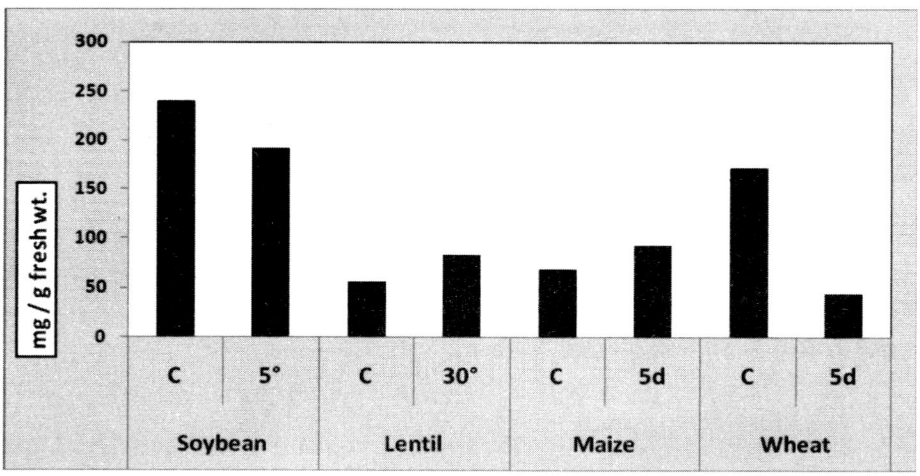

Figure 2 : Proline content in leaf & root: Soybean (cold stress) -control (25°C), Treated (5°C) Lentil (heat stress) -control (19°C), Treated (30°C) Maize & Wheat (water stress) - control, 5 day drought.

of proline in the roots of all the plants under study showed a general increase (Fig 2). The total phenol content in case of legumes showed a marked increase in both cold and heat stress whereas in case maize it showed an increase in the total phenol content but wheat showed a decrease during the 5th day of water stress (Fig 3). The change in the content of orthodihydroxy phenol following cold stress in soybean was insignificant but in case of lentil there was a significant increase owing to heat stress while both the cereals showed a remarkable decrease in the content of orthodihydroxy phenol following water stress (Fig 4).

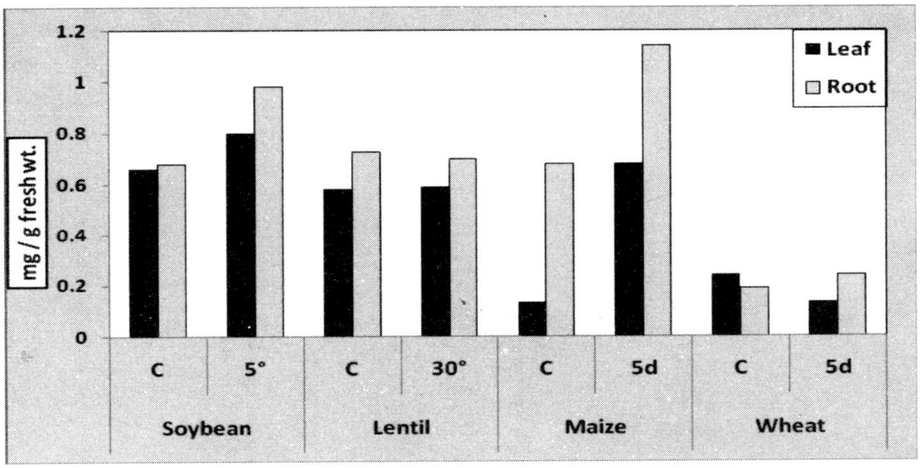

Figure 3 : Total phenol content : Soybean (cold stress) -control (25°C), Treated (5°C) Lentil (heat stress) -control (19°C), Treated (30°C) Maize & Wheat (water stress)- control, 5 day drought

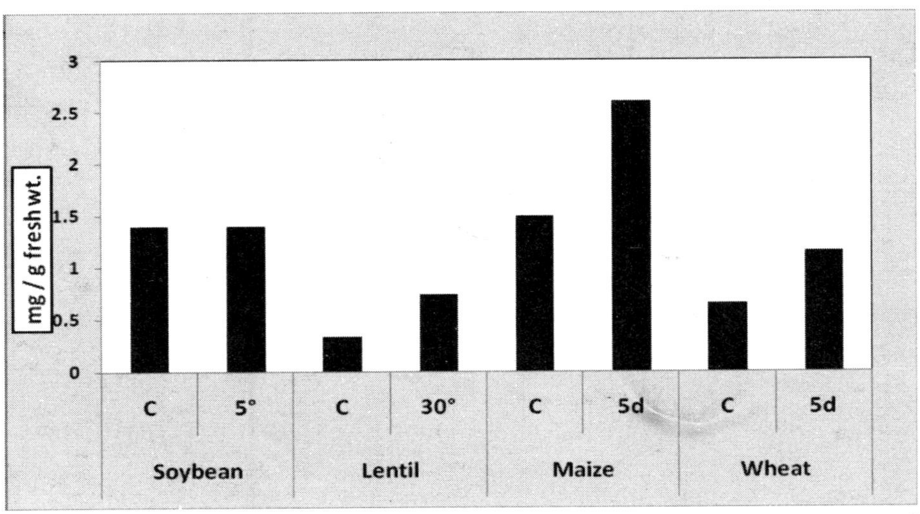

Figure 4 : Ortho dihydroxy phenol content : Soybean (cold stress) -control (25°C), Treated (5°C) Lentil (heat stress) -control (19°C), Treated (30°C) Maize & Wheat (water stress)- control, 5 day drought.

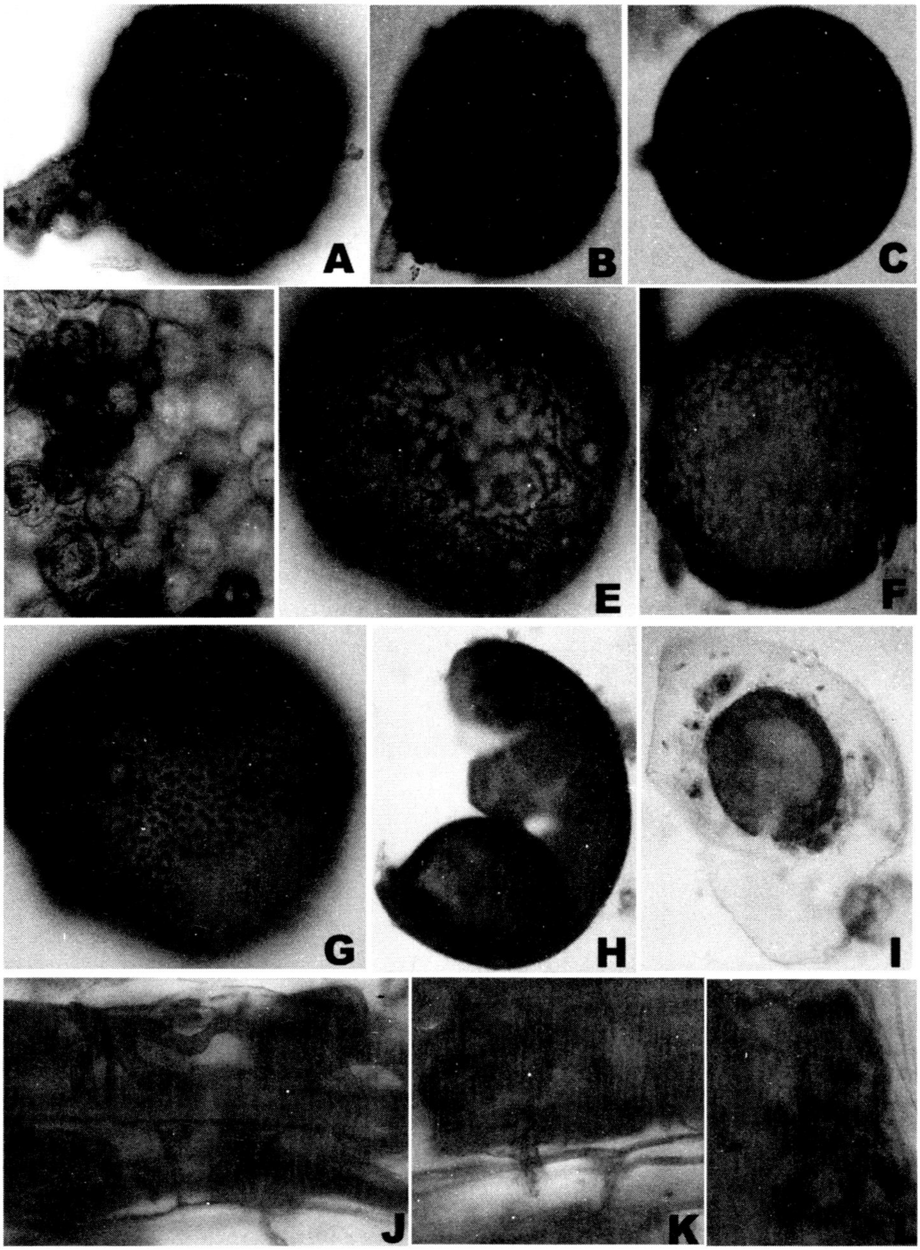

Figure 5 : A. *Glomus mossoae* ; B. *Glomus fasiculatum* ; C. *Glomus* sp. ; D. Sporocarp of *G.aggregatum* ; E. *Glomus* sp. ; F. *Acaulospora bireticulata*. G. *Acaulospora spinosa* ; H. *Scutellospora rubra* ; I. *Scutellospora* sp. J-L. Maize root cells showing arbuscules of varied strucutres.

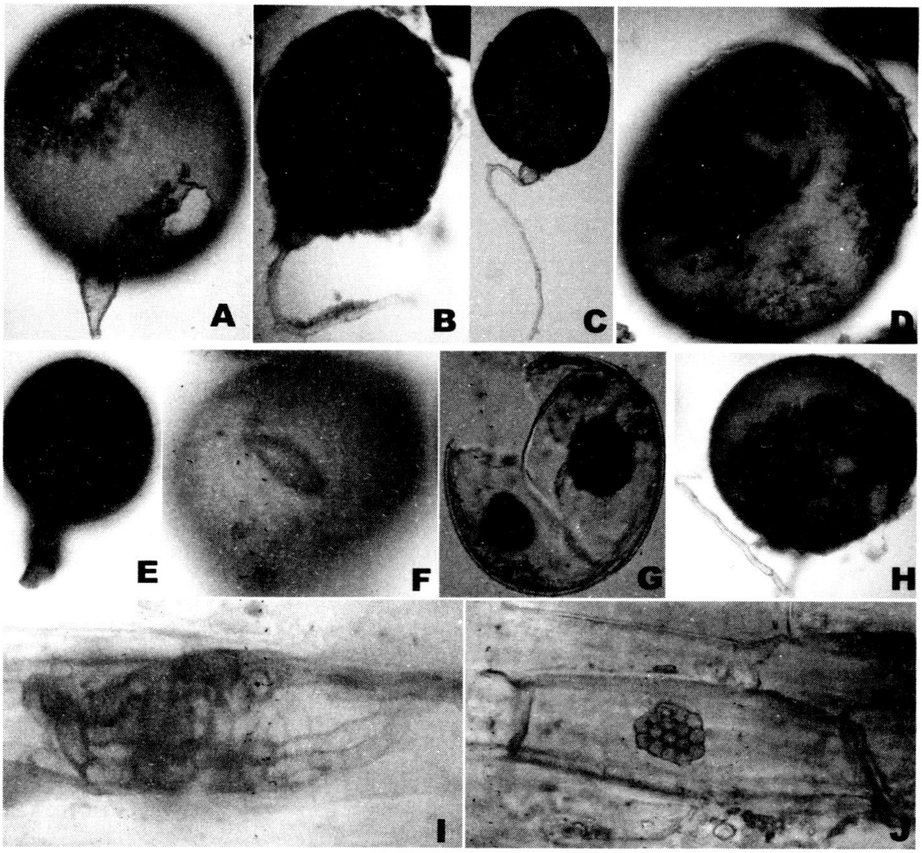

Figure 6 : A. *Gigaspora margarita* ; B. *Scutellospora rubra* ; C. *Scutellospora* sp. ; D. *Gigaspora gigantia* ; E. *Glomus constrictum* ; F. *Acaulospora* sp. ; G. *Scutellospora* sp. ; H. *Scutellospora* sp. (*Sporocarp*) ; I-J. Wheat root cells showing arbuscules of varied stuctures.

The activity of antioxidative enzymes following abiotic stresses were different in all the plants. There was a significant increase in the activity of peroxidase enzyme in all the plants following temperature and water stress, though in case of heat treatment in lentil the activity showed a decline. The maize plant showed 90% increase in the activity of peroxidase during the water stress of five days when compared to the control plant whereas there was a 50% increase in its activity in case of soybean and wheat. Ascorbate peroxidase activity declined slightly during temperature stress in soybean and lentil but the decline was not very significant in both the plants following cold and heat stress respectively, whereas cereals showed a significant increase in the activity of this enzyme owing to drought stress during the 5th day pointing to a possible role of this in dealing with drought in the cereals. There was a general increase in the activity of catalase in case of soybean, maize and wheat whereas the activity declined significantly in lentil following heat stress. Wheat showed a 50% increase in the activity when compared to control. Superoxide dismutase activity showed a decrease in all the plants. In case of soybean and maize glutathione reductase activity showed a significant increase but in wheat and lentil there was a decrease activity of this enzyme which was higher in case of wheat (Table 1).

Table 1 : Enzyme activities in plants under abiotic stresses

	Temperature stress				Water stress			
Enzyme assays	**Soybean**		**Lentil**		**Maize**		**Wheat**	
	Control (25 °C)	**Treated (5°C)**	**Control (19 °C)**	**Treated (30 °C)**	**Control**	**Treated 5 Days**	**Control**	**Treated 5 Days**
Peroxidase [a]	0.159	0.29	2.92	1.0	0.2	0.22	0.23	0.43
Ascorbate peroxi dase [b]	0.162	0.15	0.8	0.17	0.07	0.32	0.32	0.73
Catalase [c]	0.027	0.075	0.80	0.583	0.14	0.64	0.17	0.33
Superoxide dismutase [d]	0.013	0.006	0.40	0.168	0.04	0.11	0.08	0.05
Glutathione reductase [e]	0.085	0.296	0.63	0.515	0.04	0.14	0.14	0.03

a - ("A $_{460}$ mg protein $^{-1}$ min $^{-1}$);
b- ("A $_{290}$ mg protein $^{-1}$ min $^{-1}$);
c- ("A $_{245}$ mg protein $^{-1}$ min^{1});
d- (EU/mg tissue min $^{-1}$);
e- (μM NADPH oxidized mg protein^{-1} min^{-1})

Prolonged water stress showed an increase in the population of AMF significantly in both the cereals as shown in Table 2. The AMF population was found to be about 46 spores/g (approx.) in the rhizosphere of lentil and 27 in case of soybean prior to the application of stress. No significant changes occurred in AMF population after the application of temperature stress. However there was a moderate increase in the AMF population in the rhizosphere of cereals like maize and wheat under water stress for 5days. AMF spores of *Glomus* and *Gigaspora* were found to be predominant in the rhizosphere of these cereals subjected to water stress of which *Glomus fasciculatum, G. mosseae, G. aggregatum, Gigaspora gigantia, Gigaspora margarita* and few species of *Acaulospore* and *Scutellospora* were identified. Both maize and wheat plant showed significant increase in the number of AMF spores. Species of *Gigaspora* and *Scutellospora* were found to be in increased number after the of water stress than that of *Glomus* sp (Fig. 5 and 6).

Table 2 : Spore numbers of AM fungi in rhizosphere of plants following water stress

Name of AMF spores	No. of spores			
	Wheat		Maize	
	Control	Water stressed	Control	Water stressed
Glomus mosseae	22	28	15	23
G.fasciculatum	18	20	22	31
G. aggregatum	22	29	22	25
G.constrictum	09	13	30	45
Gigaspora albida	12	18	18	38
G. rosea	16	28	15	28
G. gigantea	24	37	18	41
Scutellospora sp.	13	27	16	21
Scutellospora sp.	16	24	12	19

Discussion

One month old plants of lentil and soybean were subjected to temperature stress for 4hr at 30°C and 5 °C respectively. Result of the present study also reveals the role of reactive oxygen species (ROS) scavenging antioxidant mechanism in inducing tolerance to stresses like water and temperature and AMF population during these stresses. Various enzymes involved in ROS-scavenging have been manipulated and over expressed or down regulated (Ahmad *et. al.* 2008).Peroxidases are often the first enzymes to alter their activities under stress. Peroxidase activity was increased in both the stresses. In a study with wheat genotype Gupta and Gupta (2005) reported that exposure to high temperature increased peroxidase activity following water stress in the plants. Enhanced peroxidase activity have been observed in rice seedlings under anoxia (Lee and Lin 1995). Catalases are tetrameric heme containing enzymes that catalyze the breakdown of H_2O_2 to H_2O and O_2. The effect on catalase activities vary in different plants. Unchanged CAT activities under drought stress have been observed (Smirnoff, 1993: Zhang and Kirkham 1994; Lastillo, 1996). Jiang and Huang (2001) showed the CAT activities declined under drought, heat and a combination of the two stresses. The result of the present study also revealed CAT activity decreased during high temperature stress and increased during water stress. GR activity appears to take role in lesser magnitudes of drought stress, whereas CAT and APX activities appear to be very crucial antioxidative defenses during intense stress conditions. Previous work also indicates that, the level of proline and activities of the enzymes CAT and APX are important mechanisms for the maintenance of drought tolerance in peanut plants (Akcay *et al.*, 2010). SODs are another important reactive oxygen species scavenging enzymes. Yan *et al.*, (1996) reported the reduced activity of SOD leading to accumulation of singlet oxygen in flooding stress in maize. In the present study there was a decline in SOD activity during temperature stress and water stress in both legumes and wheat but increased in case of maize. Increased SOD, POX and CAT activities in response to water stress have been reported (Srivalli *et al.* 2003, Kukreja *et al.* 2005). Our results showed that drought and temperature stress caused changes in antioxidant enzymes activity in leaves of cultivars. These results are similar in part to results obtained by Ghorbanli *et al.* (2004). Glutathione reductase activity was increased at low temperature stress in soybean and in water stress in maize but reverse is the case at high temperature stress in lentil and water stress in wheat. Jiang and Huang (2001) obtained decreased activity of GR under heat stress, which would have resulted in H_2O_2 accumulation; increased GR activity would facilitate improved stress tolerance (Tyystjarvi *et.al.*, 1999).

The rhizosphere of lentil and soybean plants analyzed after the temperature treatment showed no significant changes in the population of AM fungi. Increased activities of SOD, CAT, APOX, POX, and GR have been reported in mycorrhizal-stressed plants (Garg and Manchanda). The distribution of VAM spores during different cropping seasons was significantly different. These reports were not in agreement with others (Lopez Sanchoz and Honrubia 1990, 1992; Methew *et al.* 1990; Sulochanna and Manoharachray 19190; Vardavakis 1992) where higher counts of VAM spores were found in summer crops. This could be attributed to the prevalence of high temperature (52°C) in Haryana during summer (Dudeja and Khurana 1989) which may be lethal for VAM spores (Bhardwaj et. al 1997). It is evident that prolonged water stress enhances the AM population as the fungal hyphae penetrate more area underground to absorb water and nutrients resulting to proliferation of AM fungi.

Acknowledgement

The funding from University Grants Commission is duly acknowledged.

References

Ahmad, P., Sarwat, P. and Sharma. P. (2008) Reactive Oxygen Species, antioxidants and Signaling in Plants. *Journal of Plant Biology,* May 2008, 51 (3): 167-173

Akcay U C, Ercan O. , Kavas M , Yildiz L , Yilmaz C, Oktem H A, and Yucel M (2010): Drought-induced oxidative damage and antioxidant responses in peanut (*Arachis hypogaea* L.) seedlings .*Plant Growth Regul* (2010) 61: 21–28

Arnon D I, (1949) Copper enzyme in isolated chloroplast. Polyphenoloxidase in *Beta vulgaris* . *Plant Physiol.* 24:1-15.

Asada K, Takahashi M (1987) Production and scavenging of active oxygen in photosynthesis. In Photo inhibition, D J Kyle, C.B.Osmond and C.J. Arntzen (ed). Elsevier Science Publishers, Amsterdam 227-287

Bátes, H.S., Waldren, R.P. and Treane, D. (1973) Rapid estimation of free proline for water stress determination. *Plant soil* 39: 205-207.

Bhardwaj, S., Dudeja, S.S.and Khurana , A.L. (1997) Distribution of Vesicular – Arbuscular Mycorrhizal Fungi in the Natural Ecosystem. *Folia Microbiol.*42 (6), 589-594.

Boyler, J.S. (1982) *Plant productivity and environment science* 218: 443-498.

Bray H G and Thorpe W V (1954) Analysis of phenolic compounds of interest in metabolism. *Methods in Biochemistry Analysis* 1: 27-52

Chance B and Machly A C. (1955) Assay of Catalase and peroxidase. *Methods Enzymol* 2:764-775

Chakraborty U, Chakraborty B N and Kapoor M (1993) Changes in the level of peroxidase & Phenyl amine ammonia lyase in *Brassica napus* cultivars showing variable resistance to *Leptosphaeria maculans. Foliar Microbiol.* 38: 491-496

Daniels B A and Skipper H A. (1982) Methods for the recovery and quantitative estimation of propagules from soil. In Methods and Principles of Mycorrhizal Research. Ed. N C Schenk, American Phytopathological Society, St. Paul, Minn., pp 29-35.

Dhindsa R S, Dhindsa P L and Thorpe T A (1981) Leaf senescence: Correlated with increased level of membrane permeability lipid peroxidation and decreased superoxide dismutase and catalase. *J.Expt.Bot.*32:93-101

Dudeja, S.S., Khurana,A.L., (1989) Peristance of Bradyrhizobium sp. (Cajanus) in a sandy loam. *Soil Biol. Biochem.* 21. 709-713.

Faber BA, Zasoski RJ, Burau RG and Uriu K (1990) Zinc uptake by corn affected by vesicular-arbuscular mycorrhizae. *Plant Soil* 129: 121- 131.

Garg N and Manchanda (2008) Effect of Arbuscular Mycorrhizal Inoculation on Salt-induced Nodule Senescence in Cajanus cajan (Pigeonpea) *J Plant Growth Regul* (2008) 27:115–124

Gerdemann, J. W. and Nicolson, T. H. (1963) Spores of mycorrhizal *Endogone* extracted from soil by wet sieving and decanting. *Trans.Br. Mycol. Soc.*, 46, 235–244.

Ghorbanli, M., Ebrahimzadeh, H. and Sharifi, M. (2004) Effects of NaCl and mycorrhizal fungi on antioxidative enzymes in soybean.- *Biol. Plant.* 48: 575-581.

Gupta, S and Gupta, N.K. (2005): High temperature induced antioxidative defense mechanism in seedlings of contrasting wheat genotypers. *Indian J. Plant Physiol.* 10: 73-75.

Hare, P.D., Cress, W.A. and Van Staden, J. (1999) Proline biosynthesis and degradation: a model system for elucidating stress related signal transduction. *J. exp. Bot.* 50: 413-434.

Jakobsen I, Abbott LK and Robson AD (1992) External hyphae of vesicular-arbuscular mycorrhizal fungi associated with *Trifolium subterraneum* L. 1. Spread of hyphae and phosphorus inflow into roots. *New Phytol.* 120: 371-380.

Jiang, Y and Huang, B(2001) Effect of calcium on antioxidant activities and water relations associated with heat tolerane in two cool- season grasses. *J. Exp. Bot.* 52: 341-349.

Kothari SK, Marschner H and Romheld V (1991) Contribution of the VA mycorrhizal hyphae in acquisition of phosphorus and zinc by maize grown in calcareous soil. *Plant Soil* 131: 177-185.

Larkindale, J., Knight, M.R. (2002) Protection against heat stress induced Oxidative damage in Arabidopsis involves calcium abscisic acid, *Plant Physiol.* 128: 682-695.

Lee D H and Lee C B (2000) Chilling stress induced changes of antioxidant enzymes in the leaves of cucumber in gel enzyme activity assays. *Plant sci.* 159:75-85

Lee,T.M. and Lin, Y.N. (1995) Changes in soluble and cell wall – bound peroxidase activities with growth in anoxia – treated rice (*Oryza sativa L.*) coleoptiles and roots. *Plant Sc.*106: 1-7.

Li XL, Marschner H and Romheld V (1991) Acquisition of phosphorus and copper by VA-mycorhizal hyphae and root to shoot transport in white clover. *Plant Soil* 136: 49-57.

Lopez- Sachoz, M.E. and Honrubia, M., (1992) Seasonal variation of vesicular – Arbuscular Mycorrhizae in eroded soils form Southern Spain, Mycorrhiza 2, 33-39.S

Lowry O H, Rosebrough N J , Fair A L and Randall R J (1995)Protein measurements with Folin Phenol reagent. *J. Biol Chem.* 193:265-275

Liu A, Hamel C, Hamilton RI, Ma BL and Smith DL (2000) Acquisition of Cu, Zn, Mn and Fe by mycorrhizal maize (*Zea mays* L.) grown in soil at different P and micronutrient levels. Mycorrhiza 9: 331-336.

Mahadevan A and Sridhar R, (1982) *Methods in physiological plant pathology* 2nd edition, Sivakakmi Publication, India.

Methew, J., Neeraj, Shankar, A., Kaur, R., and Varma, A. (1990) VAM fungi from the rhizosphere of desert cacti, pp 44-46 in B.L. Jalali, H. Chand (Eds): Current Trends in Mycorrhizal Research. Haryana Agricultural University, Hisar.

Panneerselvam, (2007) Paclobutrazol enhances photosynthesis and ajmalicine production in *Catharanthus roseus*. *Process Biochem*, 42: 1566–1570

Ryan MH, Angus JF (2003) Arbuscular mycorrhizae in wheat and field pea crops on a low P soil: increased Zn uptake but no increase in P uptake or yield. *Plant Soil* 250: 225-239.

Somasundaram and Panneerselvam R., (2007) Water deficit stress mitigation by calcium chloride in *Catharanthus roseus*; effects on oxidative stress, proline metabolism and indole alkaloid accumulation. *Colloids Surf. B: Biointerfaces*, 60: 110–116

Somasundaram and Panneerselvam R., (2007) Induction of drought stress tolerance by ketoconazole in *Catharanthus roseus* is mediated by enhanced antioxidant potentials and secondary metabolite accumulation. *Colloids Surf. B: Biointerfaces*, 60: 201–206

Smirnoff, N. (1993) The role of active oxygen in the response of plants to water deficit and dessication. *New Phytol*. 125: 27-58.

Sulochana, T., Manoharachary, L., (1990) Impact of season on the distributionof VAM fungi associated with sesame, pp 16-17 in B.L. Jalali , H. Chand (Eds): Current Trends in Mycorrhizal Research . Haryana Agricultural University, Hisar.

Tyystjarvi,E., Riikonen, M., Arisi, A.C M, Kettunen, R., Jouanin, L.and Foyer, C.H.(1999) Photoinhibition of photosystem II in tobacco plants over expressing gulutathione reductase and poplars over expressing superoxide dismutase. *Physiol.Plant* 105: 405-416.

Vardavakis, E., (1992) Mycorrhizae endogonaceae and their seasonal variations in a Greek sand dune. *Pedobiologia* 36: 373-382.

Yan, B., Dai, Q, Liu, X. Huang, S. and Wang, Z. (1996) Flooding –induced membrane damage, lipid oxidation and activated oxygen generation in corn leaves. *Plant Soil*. 179: 261-268.

□□□

Chapter 15

BIOPROSPECTING OF BOTANICALS AGAINST *PHOMOPSIS AZADIRACHTAE* ON NEEM

M.K. SATEESH, T.R. LAKSHMEESHA, AND S.S. MOHAMMAD

Molecular Diagnostic Laboratory
Department of Microbiology & Biotechnology, Bangalore University,
Bangalore-560 056. Karnataka, India

Introduction

Azadirachta indica A. Juss. (Fig. 1) is widely known as neem, a tree that has proven value to humans throughout the dry tropics and subtropics. *A. indica* is a member of the Meliaceae (mahogany) family. Neem also called Margosa is an attractive broad-leaved, evergreen legendary medicinal tree. *Azadirachta indica* is derived from the Persian and means: 'The Free Tree of India'. It is grown from the southern tip of Kerala to the Himalayan hills, in tropical to subtropical regions, in semiarid to wet tropical regions, and from sea level to about 700 m elevation. Neem is well-known for its durable, insecticidal and bio-medical properties. This native tree of India is widespread in India, Pakistan, Myanmar, Sri Lanka, Thailand, Malaysia and Indonesia (Sateesh, 1998; Sateesh and Vedamurthy 2007). Neem has been introduced in more than 90 countries and established throughout the tropics and subtropics, especially in drier areas in Southeast Asia, the Pacific Islands, Australia, South and Central America, the Caribbean, sub-Saharan Africa, and the Middle East. *A. indica* has attracted worldwide prominence in recent years, owing to its wide range of medicinal properties (Sateesh 1998; Salako *et al.*, 2008).

Neem elaborates a vast array of biologically active compounds that are chemically diverse and structurally complex. Neem contains several active ingredients, and they act in different ways under different circumstances. Neem products are unique in that they are not outright killers. The main feature is that most of neem compounds are chemically similar and biogenetically derivable from tetracyclic terpenes. They affect about 350 insect species, 25 spp. of bacteria, 15 spp. of fungi, 12 spp. of nematodes, 3 viruses, 2 spp. of snails and is reported as non-toxic to humans (Koul and Wahab 2004). By 2009, several thousand researchers in at least sixty countries were studying various aspects of neem products. Neem tops the list of 2500 plant species reported to have pesticidal properties. Neem is the only plant from which the biopesticides are commercially manufactured, found effective, eco-friendly and acceptable to the farmers. Hence, this outstanding tree has been attracting the researchers, pharmacists agriculturists, environmentalists, foresters, economists, socialists, antiGMO's groups and also general public (Jeyakumar and Gupta 2002). Thus, neem tree is rightly called as doctor-plant or wonder tree of India or village pharmacy or bitter gem or organic pesticide or green gold or a tree for solving global problems. Neem is currently one of the world's most researched trees (Kaushik *et al.*, 2007).

Figure 1 : Neem trees: Healthy (left) and severely die-back affected tree (right)

Neem has been extensively used in Ayurveda, Unani and Homoeopathic medicine and has become a cynosure of modern medicine. Neem has been regarded as a wonder tree, the world over, capable of solving a large number of human and animal diseases. It has been an inherent part of both traditional and modern systems of medicine. The commercial value of neem has been known to mankind since time immemorial. Each part of the tree like the bark, seed, leaf, flower, wood, and twig has been in use in the traditional systems of medicine and now, it is increasingly being used on a commercial basis for large scale production of medicines, pharmaceuticals, cosmetics, pesticides, manures and animal feed (National Research Council 1992). Of all the plants that have proved useful to humanity, a few are distinguished by astonishing versatility. The coconut palm is one, bamboo another. In the more arid areas of India, this distinction is held by the hardy, fast-growing evergreen neem tree (Singh 1996).

India is the major producer of neem seeds in the world with annual production of about 660,000 tons. Neem is regarded as "The Wonder Tree" and "Nature's Drug Store", "Plant doctor" because its extracts have a vast pharmacological activity and are used as raw materials for pesticide, medicine and other commodities. Neem is considered to be "one of the most promising of all plants and the fact that it may eventually benefit every person on this planet". Probably no other plant yields as many strange and varied neem-products or has as many exploitable byproducts. In view of the fact that Neem originated in India and the country has a board genetic base, which needs recognition and protection. Hence, there is urgency to create awareness of neem's potential for sustainable development and environment conservation. In order to achieve this Neem foundation is making effort to declaring neem (*Azadirachta indica*) as the 'The National Tree of India' (Biswas *et al.*, 2002; Koul and Wahab 2004).

Eventhough the neem is known for its multifarious biological activities, it is infected by a species of mitosporic fungus causing die-back disease. Die-back of neem incited by *Phomopsis azadirachtae* Sateesh *et al.* is a destructive disease in India (Shankara Bhat *et al.*, 1998; Sateesh & Shankara Bhat 1999). The disease was reported for the first time in Mysore, Karnataka state, South India (Sateesh *et al*

1997) and now it is wide spread in many parts of India (Fathima *et al.*, 2004). Neem is an evergreen tree and die-back causes lose in terminal growing tips of neem trees. Hence, inflorescence will not be produced in such branches. Die-back results in almost always 100% loss of fruit production in severely infected individual trees. Spores are produced in large quantities by pycnidia and are spread by rain droplets and also by insects as the spores are produced in sticky mass (Sateesh & Shankara Bhat 1999).

At present die-back disease of neem is becoming a major threat due to destructive nature of the pathogen and spreading alarmingly in different parts of India. The continued field survey revealed that neem trees well known for their ever green canopy were declining due now showing destructive nature of die-back disease. Presently, it is considered as a major crippling disease of neem and almost always causing 100% loss of fruit production in severely infected individual trees (Sateesh and Vedamurthy 2007).

Bavistin 50% WP was found to be effective in controlling the disease (Sateesh & Shankara Bhat 1997). The most chemical pesticides are hazardous to human health and toxic to many non-target organisms. There are potential hazards associated with their manufacture, distribution and application, particularly if pesticides are misused. This disease can be controlled by using bavistin, but fungicides have hazardous effect on different components of the environment. WHO also estimated only about 0.1% of pesticides actually reach the target pests and over 99% move to ecosystem to contaminate land, water and air that are essential for biosphere (Sateesh 1998).

Biological control is an environmentally sound and effective means of reducing or mitigating pathogens or pests and pathogens or pest are effectively controlled through the use of natural enemies (Bettiol 1996). Meeting the goal of improving plant disease control for efficient and sustainable production systems will require an increased reliance on biologically-based pest controls.

Every year, plant pathogens are responsible for the loss of 10 to 20 percent of agricultural production worldwide despite the several billion dollars spent controlling them with synthetic chemicals. While synthetic chemicals have their place in disease control, there is a growing awareness that integrated pest management (IPM) strategies provide more environmentally sound and economically viable alternatives for agriculture. Thus, biological controls of plant pathogens have been identified as integral components of IPM strategies. However, biocontrol by using microbes for tree species is not an economically feasible method. Contrary to the detrimental synthetic pesticides, plant based antimicrobials are better alternatives as they are known to contain least environmental impact and danger to the biotic components. Plants have been an indispensable source of natural products for relief from varieties of illness for many years. Systematic screening of medicinal plants for antimicrobial activities and phytochemicals is important for finding potential new compounds for antimicrobial use (Duraipandiyan *et al.*, 2006). Today more than 85,000 botanicals have been documented for medical use globally. The world health organization estimates almost 80% of the world population rely on herbal remedies (Liu and Wang 2008). Owing to these biohazardous effects of synthetic fungicides (Heywood 1995), several fungal and bacterial biocontrol species were screened against *P. azadirachtae* for the antagonistic activities (Sateesh 1998). The present research communication reports the results of a survey that was done for antimicrobial activity against *P. azadirachtae*.

Materials and Methods

Collection of Plant Material and its preparation :Plants were screened for contamination by other pathogenic species and thoroughly checked for infection free plants. Mature leaves of healthy 25 plants (Table 1) were collected from Mysore, Karnataka, India during summer months (March-May). The Flora of Presidency of Madras and The Flora of Karnataka were used for identification and authentication of the plants. Collected plant material was washed thoroughly in running tap water, rinsed in distilled water and then dried under shade at 25 ± 2°C for about 10 days. The dried plant samples were ground well into a fine powder in a mixer grinder and sieved to give particle size of 50–150 mm. The plant powder was stored in air sealed polythene bags at room temperature before extraction.

Table 1 : Extracts of different plant species used against *Phomopsis azadirachtae* Sateesh *et al.* for antifungal activities

S.No.	Plant name	Family
01	*Aegle marmelos* (L.) Correa	Rutaceae
02	*Allium sativum* L.	Liliaceae
03	*Aloe vera* (L) Brum.	Liliaceae
04	*Asparagus officinalis* L.	Liliaceae
05	*Azadirachta indica* A. Juss.	Meliaceae
06	*Bambusa arundinacea* (Retz.) Roxb.	Poaceae
07	*Catharanthus roseus* (L.) G. Don.	Apocynaceae
08	*Centella asiatica* (L.) Urb.	Apiaceae
09	*Clerodendrum inerme* (L.) Gaertn.	Verbenaceae
10	*Coleus aromaticus* Benth.	Lamiaceae
11	*Emblica officinalis* Gaertn.	Euphorbiaceae
12	*Euphorbia pulcherrima* Klotz.	Euphorbiaceae
13	*Hibiscus vitrifolius* L.	Malvaceae
14	*Lantana camera* L.	Verbenaceae
15	*Lawsonia inermis* L.	Lythraceae
16	*Macrosolen parasiticus* (L.) Danser	Loranthaceae
17	*Moringa oleifera* Lam.	Moringaceae
18	*Ocimum sanctum* L.	Lamiaceae
19	*Origanum majorana* L.	Lamiaceae
20	*Oxalis corniculata* L.	Oxalidaceae
21	*Polyalthia longifolia* (Sonn.) Thw.	Annonaceae
22	*Pongamia glabra* Vent.	Fabaceae
23	*Prosopis juliflora* (Sw.) DC.	Mimosaceae
24	*Punica granatum* L	Punicaceae
25	*Tabebuia argentia* Britt.	Bignoniaceae

Preparation of Phytochemical Extracts : The powder of all the plants was suspended in double distilled water separately at the rate of one ml per gram of powder (1:1 v/w) and mixed with grinder for 20 minutes. Each extract was passed through a two-layered cheese cloth and then filtered through the Whatman No.1 filter paper (Dhingra and Sinclair, 1995). The filtrate was then centrifuged at 5000 rpm for 30 minutes. The supernatant was decanted and this formed standard 100% plant extract mother solution. The extract was used immediately or stored in refrigerator at 4°C and used within two days.

Screening of plant extracts for antifungal activities

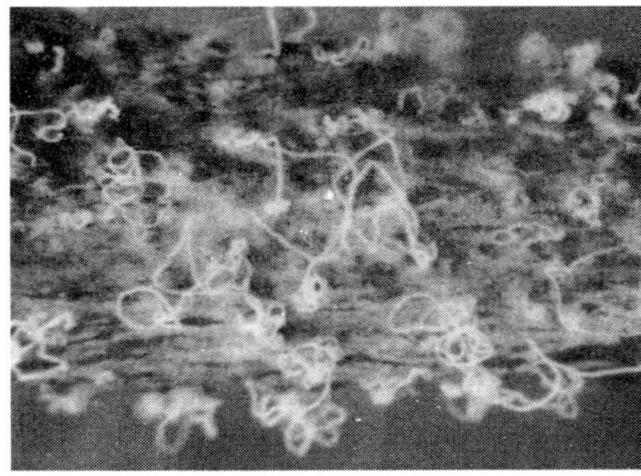

Figure 2 : Infected neem twig showing sporulation of *P. azadirachtae* Sateesh *et al.*

The evaluation of antifungal effect of plant extracts on growth of *P. azadirachtae* (Fig. 2) mycelium was studied by food-poisoning technique (Dhingra and Sinclair, 1995). Oils of two plant species *viz.*, neem and garlic were also used in this study. PDA containing five, 10, 15, 20 and 50% of the extract in the medium were prepared for each extract. The PDA medium without any plant extract served as control. The prepared plant extract amended medium was poured into sterile Petri dishes (nine cm diam.). The plant extracts were added to PDA medium before sterilization in one set and in another set plant extracts were added to sterilized PDA medium after passing through the bacteriological membrane filter. Twenty ml of the medium was poured to each Petri dish. The experiment contained five replications and was repeated four times. The mycelia agar disc (five mm diam.) of *P. azadirachtae* obtained from the margin of seven-day-old culture was used as inoculum. Each disc was inoculated at the centre of the Petri dish to both control and plant extract amended PDA medium. The colony diameter was measured after seven days of incubation at RT (25±2°C). Sporulation was observed after 15 days of incubation.

Results and Discussion

The data given in Table 2 indicate that *Lawsonia, Asparagus, Bambusa, Lantana* and *Macrosolen* extracts were effective at lower concentration *i.e.* at 20%, when compared to the other plant extracts. Fifteen per cent *Lawsonia* extract completely inhibited mycelial growth (Fig. 3). The remaining plant extracts were not effective at the same concentration. However, complete suppression of mycelial growth was observed at 50% concentration of those plant extracts. *Asparagus, Bambusa, Lantana* and *Macrosolen* inhibited only mycelial growth of *P. azadirachtae*, but did not inhibit the sporulation

(Table 3). But *Lawsonia* completely inhibited sporulation even after 20 days of incubation. *Origanum, Ocimum, Prosopis, Euphorbia, Azadirachta indica* plant extracts showed enhanced effect on growth of *P. azadirachtae*. In this case *P. azadirachtae* covered entire Petri dish at the end of six days where as in control it took eight days. Extracts from rest of the plants did not show significant effect. Both heat sterilized and membrane sterilized plant extracts showed similar results.

Table 2 : Antifungal activity of various plant species against Phomopsis azadirachtae Sateesh *et al.* growth

S.No.	Plant name	5%	10%	20%	50%
01	*Aegle marmelos* (L.) Correa	"	"	"	+
02	*Allium sativum* L.	"	"	"	+
03	*Aloe vera* (L) Brum.	"	"	"	"
04	*Asparagus officinalis* L.	"	"	"	+
05	*Azadirachta indica* A. Juss.	"	"	"	"
06	*Bambusa arundinacea* (Retz.) Roxb.	"	"	+	+
07	*Catharanthus roseus* (L.) G. Don.	"	"	"	+
08	*Centella asiatica* (L.) Urb.	"	"	"	"
09	*Clerodendrum inerme* (L.) Gaertn.	"	"	"	"
10	*Coleus aromaticus* Benth.	"	"	"	"
11	*Emblica officinalis* Gaertn.	"	"	"	"
12	*Euphorbia pulcherrima* Klotz.	"	"	"	+
13	*Hibiscus vitrifolius* L.	"	"	"	"
14	*Lantana camera* L.	"	"	+	+
15	*Lawsonia inermis* L.	+	+	+	+
16	*Macrosolen parasiticus* (L.) Danser	"	+	+	+
17	*Moringa oleifera* Lam.	"	"	"	"
18	*Ocimum sanctum* L.	"	"	"	"
19	*Origanum majorana* L.	"	"	"	"
20	*Oxalis corniculata* L.	"	"	"	"
21	*Polyalthia longifolia* (Sonn.) Thw.	"	"	"	"
22	*Pongamia glabra* Vent.	"	"	"	+
23	*Prosopis juliflora* (Sw.) DC.	"	"	"	"
24	*Punica granatum* L.	"	"	"	"
25	*Tabebuia argentia* Britt.	"	"	"	"

-No inhibition + Inhibition

Table 3 : Effects of different plant extracts (at 15% concentration) on the growth and sporulation of *Phomopsis azadirachtae* Sateesh *et al.*

S No.	Plant Name	Colonly Diameter (cm±S.E.)	Inhibition (%±S.E.)	Sporulation
01	Control	8.00 ± 0.00	000.00 ± 0.00 ^{a}	+
02	*Lantana camera* L.	5.44 ± 0.07	32.22 ± 0.07 ^{b}	+
03	*Bambusa arundinacea* (R.) Rox.	4.53 ± 0.03	43.33 ± 0.72 ^{c}	+
04	*Asparagus officinalis* L.	4.09 ± 0.05	49.56 ± 1.08 ^{d}	+
05	*Macrosolen parasiticus* (L.) D.	3.10 ± 0.11	61.25 ± 1.44 ^{e}	+
06	*Lawsonia inermis* L.0.0	± 0.00	100.0± 0.00 f	-

- Absent + Present
Values are means of three experiments ± S.E

Figures followed by different letters differ significantly (P = 0.05) when subjected Duncan's multiple range test.

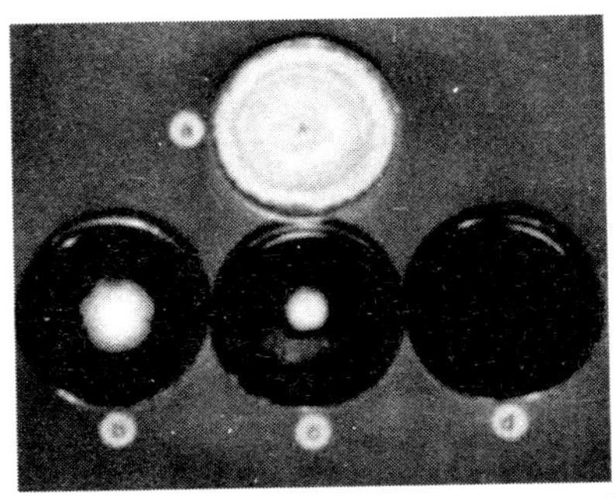

Figure 3 : Antifungal activity of *Lawsonia inermis* extract on *P. azadirachtae* Sateesh *et al.* growth a) control, b) 5% c) 10% d)15%

Neem is a very famous tree of India with worldwide fame and having magical properties known for its miraculous medicinal values from past 5000 years. This ecofriendly, native tree of India is affected by hitherto unknown species, *Phomopsis azadirachtae* Sateesh *et al.* and causing die-back disease of neem (Sateesh *et al.*, 1997). In the present day constraints on plant disease control practices, especially those imposed on the use of pesticides, biological control is increasingly capturing the imagination of many plant pathologists. Biological control can become one of the basic components in environmentally acceptable integrated pest management practices (Sandhu and Heinrich 2005; Babu and Subhasree 2009). Now world health organization banned many agriculturally important pesticides because of their broad range of toxicity for non-target organisms including humans and because of their environmental pollution (Barnard *et al.*, 1997). As neem itself an ecofriendly tree, using synthetic fungicides against the pathogen of neem is not an ecological sound practice. Green plants represent a huge reservoir of effective chemoprotectants and can provide valuable sources of effective natural pesticides (Verma and Dubey 1997; Mahajan and Das 2003). Control of *P. azadirachtae* by botanical pesticides serves as an important approach in integrated die-back disease management. The current thrust in plant protection is one promoting integrated pest management (IPM), which is ecologically sound, economically viable and socially acceptable.

The present investigations revealed that, only five plant species have given promising results against the growth of *P. azadirachtae* at lower concentration (*i.e.* at 15%) than to the other four plant species which were toxic at higher concentration. Thus *L. inermis* is most potent against the die-back disease pathogen. *L. inermis* was reported to have antifungal activity against pathogenic fungal species *viz. Pythium aphanidermatum, Phytophthora parasitica* var. *nicotianae, Colletotrichum tabacum, Alternaria alternate* on *Oryza sativa* (Natarajan and Lalitha Kumari, 1987; Satish *et al.*, 2007; Habbal *et al* 2007; *Yasmin et al., 2008.*). The present result further confirmed the earlier reports that *L. inermis* contains antimycotic secondary metabolites in its leaves (Habbal *et al.*, 2007).

Other plant extracts used in the present work have antifungal activities as reported by many precious researchers (Satish *et al.*, 2007; Zavala *et al.*, 1997). They include *Ocimum sanctum* against *Pyricularia oryzae, Cochliobolus miyabeans, and Rhizoctonia solani* (Tewari and Nayak, 1991), *Lawsonia* species against *Aspergillus flavous* and *A. parasiticus* (Sinha, 1990), *Allium sativum* against *A. parasiticus, A. ochraceus, A. versicolor* (Hitokoto *et al.*, 1978) and *Azadirachta indica* against *A. flavus* and *A. parasiticus.* But, these plants did not show such an effect against *P. azadirachtae*. The differential activity seems to be due to difference in the target organisms in both the studies. Even though certain extracts suppressed the growth of *P. azadirachtae* in one or the other concentrations, they are not economical for practical applications, because they inhibit the pathogen growth only at higher concentrations than *L. inermis.*

Toxic properties in *L. inermis* were observed only in aqueous extracts in the present study. The antifungal activity may be due to lawsone, a high phenol content in its leaves (Natarajan and Lalitha Kumari, 1987; Vaijayanthimala *et al.*, 2000). Ethanol and methanol extracts of *L. inermis* were not as effective as that of water extract against *P. azadirachtae*. Natarajan and Lalitha Kumari (1987) reported that ethanol and methanol extracts were highly effective against *Drechslera oryzae*. The contrasting result may be due to difference in the fungal species in both the studies.

Bioprospecting of naturally occurring chemicals from plants, which retards the growth and reproduction of undesirable microorganisms, would be a more sensible, realistic and ecologically sound method of future commercial pesticides for plant protection strategies, with special reference to the management of plant diseases (Gottieb *et al.*, 2002). Compared with chemical pesticides, bio-pesticides have most important advantages. They are target specific, do not destroy beneficial organisms and are bio-degradable in nature. The findings of the present research are an important step towards ecofriendly strategies for antifungal activity against *P. azadirachtae*. *L. inermis* produces a water soluble and easily extractable fungicidal compound. *L. inermis* is available throughout the state of Karnataka. And the active compounds are soluble in water, when compared to other plant species wherein they are soluble in organic solvents. In this case expensive organic solvents are not required for extraction. Thus, it is highly significant for considering *L. inermis* as a potent botanical pesticide against *P. azadirachtae*.

This helps in developing plant based antifungal compound that are used for management of disease in environmentally safe, acceptable and development of commercial formulation of botanicals. Finally it was observed that active ingredient is polar in nature and more water soluble.

References

Babu PD and Subhasree RS. (2009). Antimicrobial activities of *Lawsonia inermis* -A Review. *Academic Journal of Plant Sciences* 2(4): 231-232.

Barnard C, Padgitt M and Uri ND. (1997). Pesticide use and its measurement. *International Pest Control* 39: 161-164.

Bettiol W. (1996). Biological control of plant pathogens in Brazil: application and current research. *World Journal of Microbiology and Biotechnology*. 12: 505-510.

Biswas K, Chattopadhyay I, Banerjee RK and Bandyopadhyay U. (2002). Biological activities and medicinal properties of neem (Azadirachta indica). *Current Science* 82: 1336-1345.

Dhingra OD and Sinclair JB. (1995). Basic plant pathology methods. CRC Press, Boca Raton.

Duraipandiyan V, Ayyanar M and Ignacimuthu S. (2006). Antimicrobial activity of some ethnomedicinal plants used by Paliyar tribe from Tamil Nadu, India. *BMC Complementary and Alternative Medicine* 6: 35-42.

Fathima SN, Shankara Bhat S and Sateesh MK. (2004). Histopathology of neem seeds naturally infected with *Phomopsis azadirachtae*. *Seed Research* 32(1): 89-91.

Gottieb OR, Borin MR and Brito NR. (2002). Integration of ethnobotany and phytochemistry: dream or reality? *Phytochemistry* 60: 145-152.

Habbal OA, Al-Jabri AA and El-Hag AG. (2007). Antimicrobial properties of *Lawsonia inermis* (henna): a review. *Australian Journal of Medical Herbalism* 2(3): 10-20.

Heywood VH (1995). Global Biodiversity Assessment. UNEP, Cambridge University Press, U.K.

Hitokoto H, Morozumi S, Wauke T, Sakai S and Ueno I. (1978). Inhibitory effects of condiments and herbal drugs on the growth and toxin production of toxigenic fungi. *Mycopathologia* 66: 161-167.

Jeyakumar P and Gupta GP. (2002). Utilization of NEEM and BT for managing bolloworms in cotton. *Indian journal of entomology* 64(4): 424-433.

Kaushik NB, Singh G, Tomar, UK, Naik SN, Vir S, Bisla SS, Sharma KK, Banerjee SK and Thakkar P. (2007). Regional and habitat variability in azadirachtin content of Indian neem (*Azadirachta indica* A. Jusieu). *Current Science* 92: 212-219.

Koul O and Wahab S. (2004). Neem: Today and in the New Millennium. Springer Kluwer Academic Publisher, Netherlands. pp 276.

Liu Y and Wang MW. (2008). Botanical drugs: Challenges and opportunities: Contribution to Linnaeus Memorial Symposium 2007. *Life Sciences* 82: 445-449.

Natarajan MR and Lalitha Kumari D. (1987). Leaf extracts of antifungal agent. *Current Science* 56: 1021-1022.

Salako EA, Anjorin ST, Garba CDand Omolohunn EB. (2008). A review of neem biopesticide utilization and challenges in Central Northern Nigeria. *African Journal of Biotechnology* 7 (25): 4758–4764.

Sandhu DS and Heinrich M. (2005). The use of health foods, spices and other botanicals in the Sikh community in London. *Phytotherapy Research* 19:633-42.

Sateesh MK and Vedamurthy AB. (2007). Eco-friendly approach for control of doctor plant's (neem) die-back disease. National conference on medicinal and aromatic plants. Dec.10-12, (2007). Gulbarga University, Gulbarga.

Sateesh MK and Shankara Bhat S. (1997). *In vitro* evaluation of systemic fungicides against *Phomopsis azadirachtae* on neem. Annual meeting and Symposium of Economically Important Diseases of Crop Plants. Indian Phytopathological Society - South zone. IIHR, Bangalore.

Sateesh MK and Shankara Bhat S. (1999). Detection of see-borne *Phomopsis azadirachtae* and its transmission in *Azadirachta indica* (Neem). *Seed Science and Technology* 27: 753-759.

Sateesh MK, Shankara Bhat S and Devaki NS. (1997). *Phomopsis azadirachtae sp. nov.* from India. *Mycotaxon* 65: 517-520.

Sateesh MK. (1998). Microbiological investigations on die-back disease of neem (*Azadirachta indica* A. Juss.). Ph.D. Thesis. Mysore University, Mysore.

Satish S, Mohana DC, Raghavendra MP and Raveesha KA. (2007). Antifungal activity of some plant extracts against important seed-borne pathogens of *Aspergillus* sp. *Journal of Agricultural technology* 3(1): 109-119.

Shankara Bhat S, Sateesh MK and Devaki NS. (1998). A new destructive disease of *Azadirachta indica* incited by *Phomopsis azadirachtae*. *Current Science* 74: 17-19.

Singh, RP, Chari, MS, Raheja AK and Kraus W. (1996). *Neem and Environment*, Oxford & IBH Publishing, New Delhi. Vols. I and II, pp. 1-1198.

Sinha KK. (1990). Prevention of aflatoxin production in some cereals and oil-seeds by natural plant constitutents. *Irish Journal of Food Science and Technology* 14: 109-120.

Tewari SN and Nayak M. (1991). Activity of four plant leaf extracts against thre fungal pathogen of rice. *Tropical Agriculture* 68: 373-375.

Vaijayanthimala J, Anandi C, Udhaya V and Pugalendi KV. (2000). Anticandidal activity of certain South Indian medicinal plants. *Phytotherapy Research* 14 (3): 207-209.

Verma J and Dubey NK. (1999). Prospective of botanical and microbial products as pesticides of tomorrow. *Current Science* 76: 172-179.

Yasmin M, Hossain KS and Bashar MA. (2008). Effect of some angiospermic plants extracts on in vitro vegetative growth of Fusarium moniliforme. *Bangladesh Journal of Botany* 37(1) : 85-88.

Zavala MA, Perez S and Perez RM. (1997). Antimicrobial screening of some medicinal plants. *Phytotherapy Research* 11: 368–371.

□□□

Chapter 16

STUDIES ON POLYGALACTURONASE PRODUCTION BY *FUSARIUM OXYSPORUM* F.SP. *LYCOPERSICI* ISOLATED FROM TOMATO PLANTS (*LYCOPERSICON ESCULENTUM*)

D. NIRMALA DEVI AND C. SRINIVAS

Department of studies in Microbiology & Biotechnology, Jnana Bharathi Campus, Bangalore University, Bangalore-560 056

Introduction

Decomposition of plant cell walls during tissue invasion and pathogenesis is a characteristic feature of numerous plant diseases caused by phytopathogenic fungi. Among the wide array of cell wall-degrading enzymes are polygalacturonases (PGs), which are involved in the degradation of pectin, a complex polymer primarily found in the middle lamella and primary cell wall of higher plants. Pectin degradation plays an important role in plant diseases. Depolymerization of Pectin has been reported to be a pre requisite for further cell wall breakdown. Pectin is composed of a-1, 4-linked galacturonic acid and galacturonic acid methyl ester. Basically three types of pectinases are produced, Polygalacturonase, Pectin methylesterase and pectin lyase (Wang and Keen, 1970). Polygalacturonases are hydrolytic depolymerases with endo and exo activities. Pectin degrading enzymes including endo-polygalacturonase are among the first to be secreted during fungal infection. Besides plant cell wall degradation, the implication of pectinolytic enzymes in pathogenesis also includes the release of oligogalacturonides that can act as elicitors of the plant defense response (Hahn 1981).

F. oxysporum f.sp. *lycopersisci* (Fol) the casual agent of vascular wilt of tomato produces an array of cellwall degrading enzymes. Three distinct pectinolytic enzymes have been purified from *F. oxysporum* f.sp. *lycopersisci*, a pectate lyase and two Polygalacturonases (Di Pietro 1996). This study describes the screening, production and partial purification of extracellular Polygalacturonase from *F. oxysporum* f.sp. *lycopersisci*.

Materials and Methods

Isolation: *F. oxysporum* f.sp. *lycopersisci* was isolated from wilted tomato plants and soil samples of diseased tomato fields in Karnataka. Pure cultures of the isolates were maintained on Potato Dextrose Agar and stored as stock cultures.

Qualitative plate screening assay for Pectinase production: Preliminary screening of the isolates for pectinase production was performed using medium with 0.5% Pectin, 0.1% Yeast extract and 1.5% Agar (Hyuk Woo Kwon *et al.*, 2007). 5mm disc from a 7 day old culture of the isolates was inoculated onto petridishes containing this medium and incubated for 4 days. The plates were flooded with 1% aqueous solution of hexadecyltrimethylammoniumbromide and checked for the formation of clear zones around the colonies. 25 isolates were screened using this method (Abdel raheem *et al.*, 2002).

Enzyme production in liquid medium: Ten isolates which showed positive result in plate screening assay were selected. These isolates were grown in a synthetic medium (SM) containing 0.2g $MgSO_4.7H_2O$, 0.4g KH_2PO_4, 0.2g KCl, 1g NH_4NO_3, 0.01g $FeSO_4$, 0.01g $ZnSO_4$, 0.01g $MnSO_4$ per liter. Cultures were supplemented with 1% Pectin. The flasks were maintained at 120rpm on a shaker for 5 days (Garcia *et al.*, 1997).

Preparation of crude enzyme: Mycelia from the broth were removed by filtration through Whatman 1 filter paper. Cold acetone was added to 15ml of clarified supernatant upto 70% and centrifuged (Ibiam *et al.*, 2007). The precipitate was dissolved in 0.1M acetate buffer (pH 4.2) and stored at 4°C.

Enzyme Assay: PG activity of the culture filtrates and the crude enzyme were determined by measuring the release of reducing groups by 3, 5- dinitrosalycilic acid (DNS) method (Miller, 1959). The reaction mixture contained equal volumes of the culture filtrate and substrate (0.1% in 0.1M acetate buffer pH 4.2) incubated at 45°C for 30 minutes. The reaction was stopped by adding 0.5ml DNS, the tubes were boiled in water bath for 15mins and the absorbance read at 575nm.The PG activity of crude enzyme was determined similarly using 100ml of the enzyme preparation. Appropriate control with boiled enzyme samples was run simultaneously. The quantity of reducing sugar released was calculated from standards of D-galacturonic acid. Enzyme activity was expressed as the amount of reducing sugar released per ml per minute under the above conditions. The specific activity was calculated.

Protein determination: Protein content of the culture filtrates and the crude enzyme was determined according to the method of Lowry *et al.* using Bovine Serum Albumin (BSA) as standard (Lowry *et al.*, 1951).

Growth study: The growth of the different isolates in Synthetic medium was studied in terms of dry weight (Ervin taskin *et al.*, 2008).

Determination of molecular weight of the protein: Polyacrylamide gel electrophoresis (SDS PAGE) was carried out in slab gels composed of 5% stacking gel and 12% resolving gel. Molecular mass markers were used to determine the molecular weight of the enzyme (Garcia *et al.*, 1997).

Results and Discussions

Screening of isolates: Out of the twenty-five isolates screened by plate screening method for pectinase enzyme production, 23 isolates showed positive result with the zone size ranging between 1mm and 3mm while Fol-14 and Fol-20 were negative showing no zones around the colonies. (Table 1).

Table 1 : Plate screening assay of *F. oxysporum* f.sp. *lycopersisci* isolates for pectinase production.

Fusarium Isolates	Zone of Clearance(mm)
Fol-01	2
Fol- 02	1
Fol-03	1
Fol-04	1
Fol-05	1
Fol-06	1
Fol-07	2
Fol-08	1
Fol-09	3
Fol-10	1
Fol-11	2
Fol-12	2
Fol-13	1
Fol-14	0
Fol-15	2
Fol-16	1
Fol-17	1
Fol-18	2
Fol-19	1
Fol-20	0
Fol-21	2
Fol-22	2
Fol-23	1
Fol-24	1
Fol-25	1

Enzyme assay: Ten isolates which showed maximum zone size were selected and cultured in SM for the production of PG. The enzyme activity of the culture filtrates and the crude enzyme was determined by DNS method. The specific activity was expressed as mg of galacturonic acid per ml per milligram protein. The culture filtrate and crude enzyme of Fol-09 showed maximum specific activity of 4530.35mg/ml/mg and 6421mg/ml/mg protein respectively followed by Fol-07 which was 4094.86mg/ml/mg and 6100mg/ml/mg protein respectively. Least activity was shown by Fol-15 and Fol-13 which were 1129.58 and 1428.64 μg/ml/mg respectively (Fig.1 and 2).

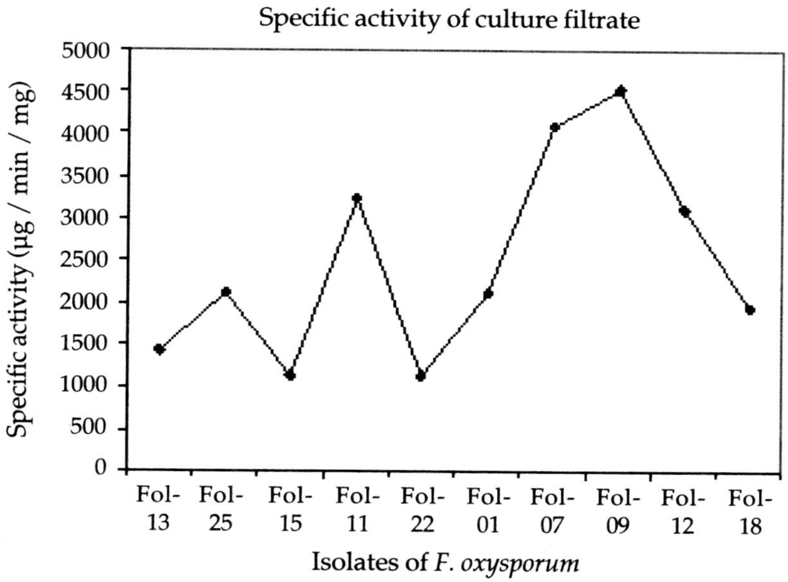

Figure 1 : Specific activity of culture filtrates of *F. oxysporum* f.sp. *lycopersisci.*

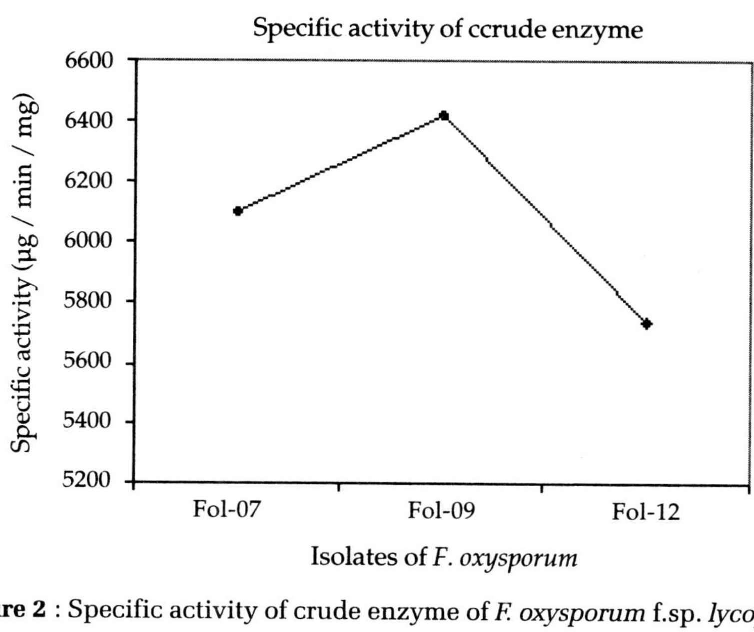

Figure 2 : Specific activity of crude enzyme of *F. oxysporum* f.sp. *lycopersisci.*

Protein Determination

The total protein content of the culture filtrates as determined by Lowry's method was found to be 74.5mg/ml in Fol-01 which was the maximum and that of Fol-07 was 72mg/ml. Crude enzymes from Fol-07 and Fol-09 showed 400 and 380 mg/ml of protein. (Fig. 3)

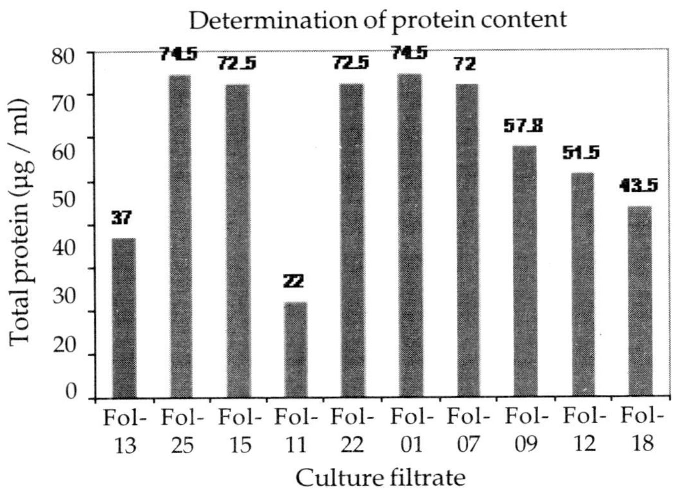

Figure 3 : Protein content of culture filtrates of *F. oxysporum* f.sp. *lycopersisci.*

Growth Study: The dry weight of the isolates grown in synthetic medium was determined, which ranged from a highest of 0.32g to a lowest of 0.10g. (Table 2).

Table 2 : Growth Study of *F. oxysporum* f.sp. *lycopersisci.*

Isolate	Mycelial dry weight (g)
Fol-13	0.22
Fol-25	0.30
Fol-15	0.29
Fol-11	0.24
Fol-22	0.32
Fol-01	0.10
Fol-07	0.19
Fol-09	0.10
Fol-12	0.13
Fol-18	0.24

SDS- PAGE: SDS-PAGE of the crude enzyme showed two major protein bands (Fig. 4). A 60kDa band was seen in the crude enzyme samples of Fol-07, 09 and 12 and a 35kDa band was observed in crude enzyme samples of Fol-09 and Fol-12 but not in Fol-07. The 60kDa band corresponds to an exopolygalacturonase and the 35kDa band corresponds to an endopolygalacturonase according to Isabel *et al.*, 2000.

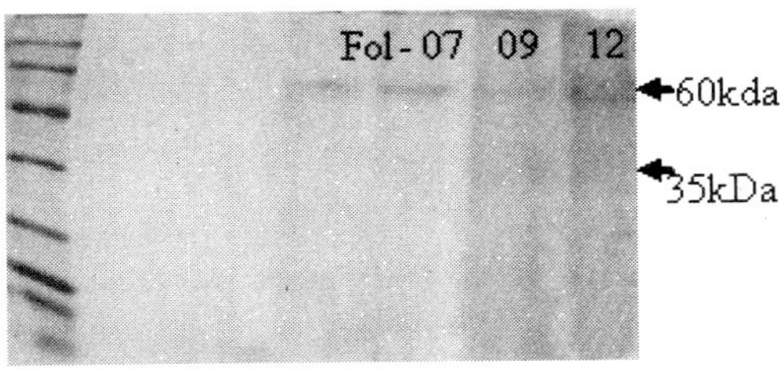

Figure 4 : SDS PAGE of crude enzyme samples. Lane 1-marker.

References

Antonio D.P., Maceira G., Huertas-Gonzales M.D.M. Carmen R.R., Zaira C., Andrea S.B.M. I and Roncero G., (1998). Endopolygalacturonase PG1 in different Formae speciales of *Fusarium oxysporum*. *Applied and Environmental Microbiology.* p. 1967-1971.

Antonio Di Pietro and M. Isabel and G. Roncero. (1996). Endopolygalacturonase from *Fusarium oxysporum f.sp. lycopersici*: Purification, Characterization and Production during infection of tomato plants. *The American Phytopathological Society*, P-1996-1015-01R.

Codner R.C. (1971). Pectinolytic and Cellulolytic enzymes in microbial modification of plant tissues. *J. appl. Bact.*34(1): 147-160.

Ernesto Favela-Torres, Tania Volke-Sepulveda and Gustava Viniegra-Gonzalez, (2006). Production of hydrolytic depolymerising pectinases. *Food Technol, Biotecnol*, 44(2): 221-227.

Ervin T., Rengin E, Erica S.S. and Vicente B.S. (2008). Screening of Aspergillus strains isolated from vineyards for pectinase production. *Journal of Food, Agriculture and Environment* Vol.6: 412-414.

Garcia M.I., Antonio D.P.M. and RonceroI G. (1997). Purification and characterization of a novel exopolygalacturonase from *Fusarium oxysporum f.sp. lycopersici .FEMS Microbiology Letters* 154: 37-43.

Hyuk W.K., Yoon J.H., Kim S.H., Hong S.B., Youngah C. and Seung J.K.(2007). Detection of extracellular enzyme activities in various Fusarium spp. *The Korean Society of Mycology*, 162-165.

Ibiam O.F.A., Arinze A.E. (2007). Determination of in vitri and in vivo production of polygalacturonase by storage mold Aspergillus niger v. Tieghem, during storage of rice seeds. *J. Appl. Sci. Environ. Manage.*: Vol. 11(2): 99-103.

Isabel G.R., Antonio D.P., Carmen R.R., Gonzales, and Maceira F.G. (2000). Role of cell wall-degrading enzymes in pathogenecity of Fusarium oxysporum. Rev Iberoam Micol, 17: S47-S53.

Miller, G.L. (1959). Anal. Biochem. 31. 426

Mohan B.G., Jyothi M., Sambhu T. and Shrestha J.K. (2007). Studies on polygalacturonase from *Aspergillus flavus*. *Scietific world*, vol 5: 19-22.

Raheem A. and Shearer C.A. (2002). Extracellular enzyme production by freshwater ascomycetes. *Fungal Diversity* II. 1-19.

□□□

Chapter 17

IMPACT OF FUNGAL METABOLITES ON THE GERMINATION STATUS OF OILSEED

R.B. KAKDE, D.P. GADGILE AND A.M. CHAVAN

Seed Pathology and Fungal Biotechnology Laboratory, Dept. of Botany, Dr. Babasaheb Ambedkar Marathwada University, Aurangabad- 431004(M.S.)

Introduction

Soybean (*Glycine max* L. Merril) has been recognized as a miracle crop due to its valuable source of high quality protein and edible oil. It contains about 44 per cent protein and 18 to 22 per cent oil. It is being used on large scale as a source of oil and protein. Safflower, (*Carthamus tinctorius* L.) is cultivated worldwide as an oilseed or ornamental crop. Their oil content varies from 24 to 36 per cent. Pathogenic fungi may attack seeds at any stage of development. Oilseeds may get infected either in field or in storage condition by pathogenic fungi. Fungi secrete their metabolites in medium during their metabolic process. These metabolites are of different types and are known to cause diseases in plants, animals and humans who eat this infected food (Schumann 1991; Singh 1991). Fungi of the genera *Aspergillus, Penicillium* and *Macrophomina* are known to produce toxic substances such as aflatoxin B_1 and B_2, Aspergillic acid, cyclopiczoic acid, ochratoxin, sterigmatocystin, which affect the health of seeds, plant and ultimately the humans who consume it. Aflatoxin inhibits the seed germination, seedling growth, root elongation, chlorophyll and carotenoid syntheses as well as protein, nucleic acid and some enzyme synthesis in soybean (Jones *et al.* 1980). The effects of such fungi on the seedlings growth include poor germination and less vigorous seedlings (Rakesh *et al.* 1984). These usually result in low yield and low income arising from poor yield quantity and quality. Several workers reported the role of toxic metabolites of storage fungi on seed germination. Soybean seed showed reduction in germination percentage, which were soaked in filtrate of *Phomopsis phaseoli* (Hilty 1988). Soybean seeds soaked in culture filtrates of *Fusarium solani, F. oxysporum, Aspergillus flavus, A. niger, Alternaria tenuis and A. alternata* for 24 hours showed reduction in percentage seed germination (Ibrahim 1987). Kakde *et al.* (2010) reported that effect of culture filtrate of *Aspergillus niger, Penicillium chrysogenum* and *Macrophomina phaseolina* of soybean seeds for different soaking period and age of culture filtrate. They found that culture filtrate of these seed-borne fungi inhibited seed germination of soybean. The present study therefore, was undertaken to evaluate the effect of toxic metabolites of storage fungi (*Aspergillus niger, Aspergillus flavus, Penicillium notatum, Macrophomina phaseolina, Fusarium oxysporum* and *Rhizopus stolonifer*) on soybean seed germination.

Materials and Methods

Storage fungi: Storage fungi (*Aspergillus niger, Penicillium notatum* and *Macrophomina phaseolina, Fusarium oxysporum and Rhizopus stolonifer*) were isolated from soybean seeds and maintained on Potato Dextrose Agar (PDA) media.

Culture Filtrate Preparation : A disc (0.5cm diameter) of mycelia and spores of each fungus was taken from the periphery of 7-day-old cultures of fungi grown on PDA medium was inoculated into 250ml conical flasks separately, each containing 100 ml of glucose nitrate broth. The broth contains (g/l): glucose 10g, potassium nitrate 2.5g, potassium dihydrogen ortho-phosphate 1g, magnesium phosphate 0.5g. The flasks were allowed incubated at room temperature for 15 days. The fungal filtrates were obtained by passing the culture through sterile Whatman No. 1 filter paper to obtain a cell-free filtrate (Umecharuba and Neachukwa, 1997).

Effect of Fungal filtrate on Seed germination: Seeds of soybean (var. JS-335) and safflower (var. Bhima) were surface sterilized with 1% Mercuric chloride solution for 1 min and rinsed several times in sterile distilled water. These oilseeds were then allowed to presoak in fungal culture filtrate for 1hr, 2hr, 3h, 4hr and 5hr. In another experiment, seeds were presoaked in different concentration (00%, 25%, 50% 75% and 100%) of culture filtrate of selected fungi. At the end of presoaking period, the seeds were removed from the filtrates and transferred into the petriplates containing two layered blotter papers soaked with sterile distilled water. About 20 seeds were sown per dish and it was then allowed to incubate for two days at room temperature. Germination counts were made after incubation period of 24h.

Results and Discussion

From Table 1 it is clear that culture filtrate of *Macrophomina phaseolina* hampered drastically the germination of soybean and safflower followed by *Aspergillus niger, Penicillium notatus, Aspergillus flavus* and *Fusarium oxysporum* as compared to other fungal filtrate. Fungal filtrate of *Rhizopus stolonifer* showed minimum effect on germination status of soybean and safflower. As compared to soybean, safflower showed maximum germination irrespective of the toxic metabolites. Table 2 shows effect of different concentrations of toxins of fungi on germination of soybean and safflower. Toxic metabolites of *Macrophomina phaseolina* inhibited the germination of soybean and safflower. Oilseed germination is lowest at 100% concentration, while it is maximum at 25% concentration. At 25% concentration of culture filtrate of *Aspergillus flavus* and *Fusarium oxysporum* germination of soybean and safflower was found to be decreased. At 50% concentration of culture filtrate of *Aspergillus niger* and *Macrophomina phaseolina* soybean seed germination was found to be hampered. Culture filtrate of *Aspergillus flavus* and *Fusarium oxysporum* decreased the germination of soybean while in case of safflower *Aspergillus niger* and *Macrophomina phaseolina* decreased the germination at 75% concentration of culture filtrate. It was observed that at 100% concentration of culture filtrate of *Penicillium notatum* and *Macrophomina phaseolina* soybean seed germination percentage was lowered; in presence of culture filtrate of *Aspergillus flavus* and *Macrophomina phaseolina* safflower seed germination percentage was found to be reduced. From the results it can be concluded that culture filtrate of *Aspergillus niger, Penicillium notatum* and *Macrophomina phaseolina* inhibited seed germination, which means that metabolites are discharged by the tested fungi in the media in which they were grown. These toxic metabolites can inhibit and reduce percentage of germination, which means that with increasing soaking period and concentration of fungal filtrate will decrease the germination of soybean and safflower. It can also be concluded that maximum soaking period and concentration of fungi will allow the maximum soaking of the toxins through seed coat, scutellum which ultimately affect on the physiology of the seed. These results are agreed with results given by Haikal (2008). He reported that toxic metabolites secreted by of *Aspergillus niger, Fusarium culmorium, Penicillium sp., Rhizoctonia solani* reduced the percentage of seed germination of soybean. Hilty and Lee (1988) found that filtrate from *Phomopsis phaseoli* reduced germination of soybean; this is due to toxins secreted in the media. Madhavrao and Thakur (1978) reported that culture filtrate of *Sclerotium rolfsii* hampered the germination of *Solanum melongena.* Sharma and Sharma (1983) found that maximum seed germination of Sweet fennel was due to *Fusarium moniliforme* and *Penicillum citrinum.* According to Sharma and Singh (2003), toxic filtrates of *Aspergillus niger, Fusarium oxysporum and Rhizoctonia solani* causes reduction in seed germination of pea. Mehan and Chohan (1974) recorded 100% inhibition in seed germination of *Phaseolus aureus* due to toxins present in culture filtrate of toxigenic strain of *Aspergillus flavus.*

Table 1 : Effect of different soaking period on germination % of Soybean and Safflower

Fungi	Soybean					Safflower				
	1hr	2hr	3hr	4hr	5hr	1hr	2hr	3hr	4hr	5hr
Aspergillus niger	18	17	15	15	09	19	18	15	12	10
Aspergillus flavus	18	17	15	13	11	19	19	17	16	13
Penicillium notatum	18	17	15	13	10	19	18	16	14	11
Fusarium oxysporum	18	16	14	13	11	18	18	17	16	14
Macrophomina phaseolina	17	15	13	11	08	19	16	15	12	09
Rhizopus stolonifer	18	16	16	14	12	19	17	16	14	10
Control	18	19	20	20	20	19	20	20	19	20

Table 2 : Effect of different concentration of fungal filtrates on germination % of Soybean and Safflower

Fungi	Soybean					Safflower				
	00%	25%	50%	75%	100%	00%	25%	50%	75%	100%
Aspergillus niger	20	19	14	11	08	20	20	18	13	11
Aspergillus flavus	19	13	15	10	08	20	19	17	15	09
Penicillium notatum	20	19	17	13	05	20	19	16	14	10
Fusarium oxysporum	19	14	12	08	06	20	18	17	15	11
Macrophomina phaseolina	20	16	10	10	04	20	20	15	10	08
Rhizopus stolonifer	19	17	15	12	09	19	18	16	17	11
Control	20	19	20	20	20	20	20	19	20	20

Acknowledgement

First and foremost authors are grateful to UGC, New Delhi for giving financial support in the form of UGC Major Research Project. Authors are also thankful to Professor and Head Dept of Botany, Dr. Babasaheb Ambedkar Marathwada University, Aurangabad for giving research facilities.

References

Hilty TC and Lee HL. (1988). Hot acidified zinc sulphate as seed soaking agent for the control of crucifer black rot. *Plant Protec Bul* 30: 245-248.

Haikal NZ. (2008). Effect of Filtrate of Pathogenic Fungi of Soybean on Seed germination and seedling parameters. *J Apl Sci* 4: 48-52.

Ibraheem SA, Okesha AM and Mlhathem KT. (1987). Interrelationship between protein and oil content of soybean seed with some associated fungi. *J Agric Water resources Res Plant Prodi* 6: 53-66.

Jones HC, Chancey JC, Morton WA, Dashek WV and Llewellyn GC. (1980). Toxic responses of germinating pollen and soybeans to aflatoxins. *Mycopathol* 72: 67–73.

Kakde RB, Gadgile DP, Pangrikar PP, Hatti AD, Gaikwad RS and Chavan AM. (2010). Soybean seed germination status under the influence of fungal metabolites. *Bionano Frontier* 3 (1): 116-117.

Madhavrao P and Thakur KS. (1978). Toxic Potentialities of Culture filtrate of Storage Fungi on the Germination of Soybean. *Ind Phytopath* 32: 39-41.

Maheshwari RK, Mathur SK and Mathur A. (1984). Effect of seed-borne fungi on seed viability and seedling vigour. *Ind Phytopath* 37: 713-714.

Mehan VK and Chohan JS. (1974). Virulence of aflatoxigenic and nontoxigenic isolates of *Apergillus flavus* Links ex Fr. on groundnut seedling. *Phytopath Medit.* 13: 108-109.

Schumann GL. (1991). Plant Disease: their biology and social impact. *The American Phytopath Soci* St. Paul, Minnesota, 377 p.

Sharma AK and Sharma KD. (1983). Effect of Fungal metabolites on germination of sweet fennel *(Foeniculum vulgare* Mill.) seeds. *Acta Pathologicai* 17: 81-84.

Sharma P and Singh SD. (2003). Effect of fungal metabolites on germination and vigour of pea. *J Mahar Agr Uni* 28: 269-270.

Singh K, Frisvad JC, Thrane U and Mathur SB. (1991). An illustrated manual on *Aspergilli, Fusaria, Penicillia* and their mycotoxins. Denmark, 133p.

Umecharuba GI and Neachukwa EO. (1997). The effect of filtrates of seed-borne fungi of African yam-bean on seed germination and seedling development. *Global J Pure and Appl Sci,* 3,165-176.

□□□

Chapter **18**

GENETIC ANALYSIS OF CARBENDAZIM - TOLERANT AND SENSITIVE ISOLATES OF *FUSARIUM OXYSPORUM* F.SP. *PISI*, THE CAUSAL AGENT OF PEA WILT

S.K. GUPTA, M. SHARMA, S. RANA AND K. JARIAL

Department of Mycology and Plant Pathology,
Dr. Y.S. Parmar University of Horticulture and Forestry, Nauni, Solan 173230

Introduction

Pea (*Pisum sativum* L.) is an important leguminous crop grown both as vegetable and pulse having 21.3 to 26.2 per cent protein. As a cool season crop, it is extensively grown in temperate zone, but restricted to cooler altitudes in the tropics and winter season in subtropics. In mid hills of Himachal Pradesh, pea is grown during September-November to March-April. Autumn crop (September to November) is severely affected with Fusarium wilt (*Fusarium oxysporum* f.sp. *pisi*) which reduces the potential yields significantly; higher soil temperature ranging between 21.8 to 22.5 °C and soil moisture between 13.0 to 14.4 per cent favour wilt incidence. Symptoms appear as yellowing of the lower leaves and a stunting or dwarfing of plant growth. Losses upto 60 per cent have been observed in the crop during years of wilt epiphytotics in Himachal Pradesh (Kumar 1993). Carbendazim has been recommended to combat the ravages of this disease but some farmers are complaining about its reduced efficiency in managing this disease. Keeping this in view, different isolates of *Fusarium oxysporum* f.sp. *pisi* were collected from different pea growing areas of the state and present studies were carried out with the following objectives: to determine the variation in cultural and morphological characteristics; pathogenic variability on cv. "Arkel" of pea and sensitivity of different isolates to carbendazim. In addition, molecular analysis using RAPD primers was carried out to resolve genetic variability of carbendazim - tolerant and carbendazim - sensitive isolates.

Material and Methods

Twenty isolates of *F. oxysporum* f.sp. *pisi* were collected from different pea growing areas of Himachal Pradesh and were designated as I_1 to I_{20}. In order to study the cultural and morphological characteristics, mycelial bits (5 mm diameter) of each isolate were taken from the actively growing cultures and centrally placed on Petri plates (90 mm) containing sterilized PDA medium. Plates were incubated at 25±1°C for 2 weeks. Data in relation to cultural characteristics and average microconidial, macroconidial, mycelial width and chlamydospores size were recorded.

Comparative sensitivity studies of different isolates were carried out with carbendazim (20 mg/l) under *in vitro* conditions by poisoned food technique (Falck 1907). Unamended PDA plates were kept as control for comparison. Each treatment was replicated thrice and incubated at 25±1°C. The data on diametric growth of isolates were recorded regularly at 24 hour interval till the control plates became full. Growth rate (rg) mm/h was calculated as given by Gupta and Upmanyu (2005):

$$\mathbf{rg = cd_2 - cd_1/t_2 - t_1}$$

Where, **rg** = Rate of mycelial growth, $\mathbf{cd_2}$ = mean diameter of colony at time $\mathbf{t_2}$, $\mathbf{cd_1}$ = mean diameter of colony at time $\mathbf{t_1}$, $\mathbf{t_2}$= time when mean diameter of colony was $\mathbf{cd_2}$, $\mathbf{t_1}$= time when mean diameter of colony was $\mathbf{cd_1}$.

In order to see the pathogenic variability among *F. oxysporum* f.sp. *pisi* isolates on cv. "Arkel" of pea, mass culture of test pathogen isolates was prepared on maize sand meal medium. Pea plants (21 days old) raised in pots were inoculated with the culture and the data on incubation period was recorded. The per cent disease incidence was calculated as:

$$\text{Disease incidence (\%)} = \frac{\text{Number of diseased plants}}{\text{Total number of plants observed}} \times 100$$

For RAPD analysis of pathogen isolates, all 20 isolates were grown in potato dextrose broth at 25$\pm$ 1° C for 14 days and mycelia were harvested by filtration through filter paper (Whatman No. 1). For DNA isolation 250 mg lyophilized mycelium of each isolate was ground in liquid nitrogen in a small sterilized mortar and total fungal genomic DNA was isolated according to CTAB (Hexa-decyl tri-methyl ammonium bromide, Sigma Chemical CO., St. Louis, USA) method (Murray and Thompson 1980). Isolated DNA was purified by using 5ml RNAase (10mg/ml) and quantified by running 2ml of each DNA sample on 1.0 per cent agarose gel to get desired concentration. Initially, forty 10-mer random primers of A and D series of Operon Technologies were tried (Table 3) for amplifications of two isolates and two primers giving maximum polymorphs was selected for further studies of all isolates. The PCR reaction was set up with a total reaction volume of 25 ml comprised of 20 ng fungal genomic DNA, 2.5 ml PCR buffer (10 X) containing $MgCl_2$ (15mM), 1 ml of 10 mM dNTP, 0.2 ml of Taq DNA polymerase (3 U/ml), and 2 ml (15 ng) each of 10-mer oligonucleotide primer (Operon Technologies). The reactions were performed in a thermocycler (Applied Boisystems 9700) set to the following: denaturation at 94°C for 2 min, 45 cycles of 94°C at 1 min, 37°C for 1 min and 2 min 72°C for 2 min. PCR amplification products were electrophoretically separated on 1 percent agarose gel stained with EtBr in TAE buffer and photographed under UV light in a Gel Documentation System (Alpha Innotech Corporation). Isolates were screened twice with primers OPAD 15 and OPAD 16. Repeatable polymorphisms were scored as 1 for presence of a band and 0 for absence. 1kb DNA ladder was loaded in two lanes on agarose gel for determining the band size. The resulting matrix was used to compute DICE's similarity coefficient for all pairwise comparisons of the isolates (Dice 1945). Cluster analysis by the unweighted pair-group method of arithmetic averages (Sneath and Sokal, 1973) were performed with the similarity values using the SAHN procedure of the program NTSYS-pc (Rohlf 1998).

Results and Discussions

Cultural and morphological variation : The mycelial colour varied from white to dull white with slightly yellowish (I_2) to pinkish tinge. The microconidial size varied from 3.16 × 3.16 μm (I_{19}) to 9.13 × 5.44 μm (I_7) whereas macroconidial size varied from 11.77 × 3.16 μm (I_{19}) to 24.60 × 5.91 μm (I_7). Mycelium width and chlamydospore dimensions also varied in all the twenty isolates of the test pathogen. Maximum mycelial width was recorded in I_2 (7.66 μm) followed by 6.67 μm in I_{12} and minimum was recorded in I_8 (2.80 μm). All isolates formed chlamydospores on PDA medium except I_2 in which no chlamydospore formation was recorded. Chlamydospore dimensions varied from 6.85 × 6.15 (I_{14}) to 13.70 × 10.18 mm(I_5). Variation in cultural and morphological characteristics of *F. oxysporum* f.sp. *pisi* isolates has earlier also been reported by Verma and Dohroo (2003).

Carbendazim resistance : Carbendazim was found to be most effective, giving 100 per cent inhibition in the mycelial growth at 25, 50, 100 and 200 ppm concentrations (Verma 2001). Isolates varied in their response to carbendazim amended media at 20 mg/l concentration (Table 1), showing mycelial growth rate variation ranging from 0.000 (I_1, I_8, I_{10}, I_{12}, I_{13}, I_{14}, I_{15}, I_{16}, I_{18}, I_{19} and I_{20}) to 0.050 mm/hr in I_{11}.

On the basis of carbendazim sensitivity, isolates were grouped into two groups. First group comprised of those isolates which were completely inhibited by carbendazim, which included I_1, I_8, I_{10}, I_{12}, I_{13}, I_{14}, I_{15}, I_{16}, I_{18}, I_{19} and I_{20} , whereas, second group comprised of those isolates which exhibited some degree of mycelial growth rate and this group included nine isolates viz., I_2, I_3, I_4, I_5 I_6, I_7, I_9, I_{11} and I_{17}. Isolates also varied in terms of mycelial growth rate on unamended PDA plates, exhibiting variation from 0.20 (I_2) to 0.34 (I_4).

Pathogenic variability : Studies on pathogenic variability (Table 2) suggested that there were significant differences in aggressiveness among the isolates. Disease incidence varied from 0.00 to 87.00 per cent. Maximum disease incidence was observed in I_{11} (87.00 %) followed by I_1 (62.50 %) and I_{20} (57.14 %), whereas, no disease incidence was observed in I_7, I_{10} and I_{12} after 7 days of inoculation of 21 days old pea seedlings in pots. The isolates were grouped into six groups depending upon the incubation period (Table 2). Importantly out of twenty isolates, eight isolates (I_2, I_5, I_8, I_{14} I_{15}, I_{16}, I_{17}, and I_{19}) exhibited incubation period of 7 days.

Table 1 : Effect of carbendazim (20 mg/l) amended PDA media on mycelial

	Carbendazim amended medium (20 mg/l)	Unamended medium	Mean
I_1 (Nauni 1)	0.000	0.310	0.155
I_2 (Pandah)	0.029	0.200	0.114
I_3 (Dharon ki Dhar)	0.037	0.280	0.158
I_4 (Gaura)	0.048	0.340	0.194
I_5 (Nohra)	0.033	0.220	0.126
I_6 (Rajgarh)	0.041	0.210	0.125
I_7 (Solan)	0.008	0.220	0.114
I_8 (Nauni 2)	0.000	0.250	0.125
I_9 (Nauni 3)	0.016	0.260	0.138
I_{10} (Nauni 4)	0.000	0.250	0.125
I_{11} (Chauras)	0.050	0.240	0.145
I_{12} (Shallana)	0.000	0.270	0.135
I_{13} (Chaug)	0.000	0.280	0.140
I_{14} (Chaugteli)	0.000	0.290	0.145
I_{15} (Kanda)	0.000	0.290	0.145
I_{16} (Shamoga 1)	0.000	0.290	0.145
I_{17} (Didag)	0.029	0.250	0.139
I_{18} (Shamoga 2)	0.000	0.300	0.150
I_{19} (Churwadhar)	0.000	0.290	0.145
I_{20} (Kullu)	0.000	0.282	0.141
$CD_{0.05}$	**0.001**	**0.015**	**0.007**

Molecular analysis : Screening of forty 10-mer oligonucleotide RAPD primers revealed all primers to yield strong and reproducible DNA amplicons (bands) of *F. oxysporum* f. sp. *pisi* isolates (Table 3). The level of polymorphism was different with different primers in two isolates. Maximum number of bands (11) was scored with primer OPAD 15 followed by 10 bands with OPAD 16. DNA fingerprints obtained with RAPD primer OPAD 15 is shown in Figure 1. Multivariate analysis was conducted to generate a similarity matrix using DICE coefficient. To estimate genetic diversity and relatedness

Table 2 : Pathogenic variability of *Fusarium oxysporum* f.sp. *pisi* isolates on cv. "Arkel" of pea (*Pisum sativum* L.)

Group(s)	Incubation period (days)	Isolate	Per cent disease incidence after 7 days
Group 1	3	I_1	62.50 (52.24)
		I_3	44.44 (41.80)
		I_4	63.63 (52.91)
		I_{18}	28.58 (32.31)
Group 2	4	I_{20}	57.14 (49.10)
Group 3	5	I_6	33.33 (35.04)
		I_9	50.00 (45.00)
Group 4	6	I_{11}	87.00 (68.90)
		I_{13}	12.50 (20.64)
Group 5	7	I_2	14.28 (22.20)
		I_5	16.67 (24.09)
		I_8	25.00 (29.92)
		I_{14}	16.67 (24.09)
		I_{15}	16.67 (24.09)
		I_{16}	11.11 (19.45)
		I_{17}	28.58 (32.29)
		I_{19}	20.22 (26.45)
Group 6	9	I_7	00.00 (0.00)
		I_{10}	00.00 (0.00)
		I_{12}	00.00 (0.00)
	$CD_{0.05}$		**2.71**

Figures in parentheses are arc sine transformed values

among nine isolates of *F. oxysporum* f sp *pisi*, a dendrogram was constructed with 200 fragments (including 103 amplicons of different isolates) obtained from RAPD data of primer OPAD 15 as shown in figure 2 by UPGMA analysis to determine grouping of isolates. The maximum coefficient (100.0) was observed for the groups of isolates 16, 17 and 20; isolates 4 and 18 and isolates 13 and 15. The dendrogram produced from computerized cluster analysis of the DNA fingerprints revealed a great deal of heterogeneity among isolates as expected after the DNA banding pattern analysis of gel image (Figure 2). All the isolates were classified in three major groups. Group I comprised of three carbendazim tolerant isolates (I_{17}, I_4 and I_9) and eight carbendazim

Table 3 : Primer, its sequence and amplification detected during preliminary studies on primer selection

S.No.	Primer	Sequence(5'to3')	Amplification	No. of amplicons
1	OPAA 1	CAGGCCCTTC	Yes	6
2	OPAA 2	TGCCGAGCTG	Yes	6
3	OPAA 3	AGTCAGCCAC	Yes	6
4	OPAA 4	AATCGGGCTG	Yes	7
5	OPAA 5	AGGGGTCTTG	Yes	5
6	OPAA 6	GGTCCCTGAC	Yes	6
7	OPAA 7	GAAACGGGTG	Yes	6
8	OPAA 8	GTGACGTAGG	Yes	4
9	OPAA 9	GGGTAACGCC	Yes	7
10	OPAA 10	GTGATCGCAG	Yes	5
11	OPAA 11	CAATCGCCGT	Yes	4
12	OPAA 12	TCGGCGATAG	Yes	4
13	OPAA 13	CAGCACCCAC	Yes	2
14	OPAA 14	TCTGTGCTGG	Yes	5
15	OPAA 15	TTCCGAACCC	Yes	3
16	OPAA 16	AGCCAGCGAA	Yes	6
17	OPAA 17	GACCGCTTGT	Yes	4
18	OPAA 18	AGGTGACCGT	Yes	5
19	OPAA 19	CAAACGTCGG	Yes	6
20	OPAA 20	GTTGCGATCC	Yes	6
21	OPAD 1	ACCGCGAAGG	Yes	8
22	OPAD 2	GGACCCAACC	Yes	5
23	OPAD 3	GTCGCCGTCA	Yes	5
24	OPAD 4	TCTGGTGAGG	Yes	4
25	OPAD 5	TGAGCGGACA	Yes	7
26	OPAD 6	ACCTGAACGG	Yes	4
27	OPAD 7	TTGGCACGGG	Yes	3
28	OPAD 8	GTGTGCCCCA	Yes	3
29	OPAD 9	CTCTGGAGAC	Yes	5
30	OPAD 10	GGTCTACACC	Yes	6
31	OPAD 11	AGCGCCATTG	Yes	9
32	OPAD 12	CACCGTATCC	Yes	4
33	OPAD 13	GGGGTGACGA	Yes	8
34	OPAD 14	CTTCCCCAAG	Yes	8
35	OPAD 15	CATCCGTGCT	Yes	11
36	OPAD 16	AGGGCGTAAG	Yes	10
37	OPAD 17	TTTCCCACGG	Yes	5
38	OPAD 18	GAGAGCCAAC	Yes	5
39	OPAD 19	CTGGGGACTT	Yes	7
40	OPAD 20	ACCCGGTCAC	Yes	8

sensitive isolates (I_1, I_{16}, I_{20}, I_{18}, I_{19}, I_{12}, I_8, and I_{10}). Isolate I_4 and I_{18}, having three days of incubation period, grouped together with 100 percent similarity. Isolates I_8 and I_9 formed a sub-group. Both of these isolates were isolated from the same region i.e., Nauni, Solan. Group II comprised of four carbendazim tolerant isolates (I_2, I_3, I_6 and I_{11}) and three sensitive isolates (I_{13}, I_{15} and I_{14}). Isolates I_2, I_{13}- and I_{15}- clustered together as a sub-group. All these isolates have incubation period of 7 days. Group III consisted of two carbendazim tolerant isolates I_5 and I_7. Groups I and II separated from each other at 44 per cent similarity coefficient value while Group III separated from rest two groups at 30 per cent similarity coefficient value.

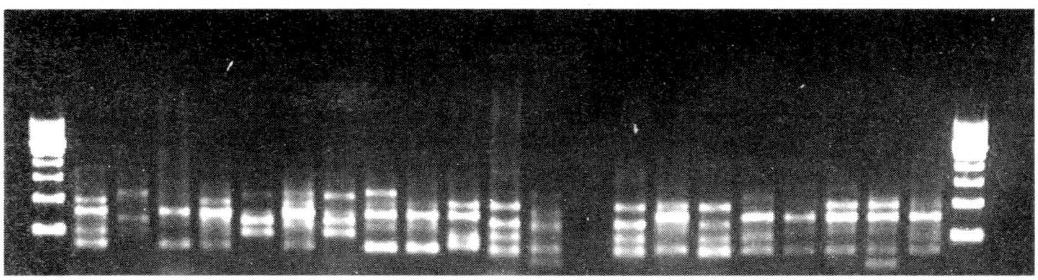

Figure 1 : Amplification products of *Fusarium oxysporum* f sp *pisi* isolates with primer OPAD-15. Lane M : 1kb ladder (molecular weight marker); Lanes 1 to 12: Isolate No 1 to 12, respectively; Lanes 14 to 21: Isolate No 13 to 20, respectively.

High degree of pathogenic variability has already been described by earlier workers in *F. oxysporum* f.sp. *pisi* (Linford 1928; Snyder 1933; Haglund and Kraft 1970; Armstrong and Armstrong 1974; Whitehead *et al.* 1992; Kraft, 1994; Stefanelli *et al.* 1996; Verma and Dohroo 2003). Bauxton (1956) suggested that *Fusarium oxysporum* has a system resembling the parasexual cycle of other fungi imperfecti, which allow different nuclei occasionally to exchange genetic material, and thus may be a factor responsible for the development of new races. Kraft and Haglund (1978) classified *F.o.* f.sp. *pisi* in 11 races. RAPD has been successfully used to identify strains and races in phytopathogenic fungi (Chiocchetti *et al.* 1999; Logrieco *et al.* 1990). It has been used for studying inter and intraspecific variability among population from different and same geographic areas (Walker *et al.* 2001). In the present studies, varied response of different isolates has been observed towards carbendazim (20 mg/l). Nine isolates exhibited some degree of mycelial growth rate showing tolerance to carbendazim.

Carbendazim being a site specific fungicide, the problem of resistance development in the target pathogen may be brought about (Thind 2008). The case of resistance development due to continuous application of benzimidazole fungicide has already been reported (Wicks 1974). The RAPD pattern visualizes variations in total DNA and thus is suitable for differentiating *F. oxysporum* f sp *pisi* isolates. RAPD results obtained in present study enabled fast variability analysis for *F. oxysporum* f sp *pisi* isolates. Group III clearly comprised of two isolates i.e., I_5 and I_{17} which were tolerant to carbendazim while Group I and Group II comprised of both tolerant and sensitive isolates. In Group II, four isolates, out of seven, were carbendazim tolerant. Two carbendazim sensitive isolates I_{13} and I_{15} clustered together with one carbendazim tolerant isolate I_2. These isolates have incubation period of 7 days which might be the reason of their clustering. Similarly in Group I carbendazim tolerant isolates I-$_4$ and I_{17} clustered with carbendazim sensitive isolates I_{18} and I-$_{16}$ respectively. These subgroups have same incubation period of 3 days (I_4 and I_{18}) and 7 days (I-$_{17}$ and I_{16}). In addition, I_8 (carbendazim sensitive) and I_9

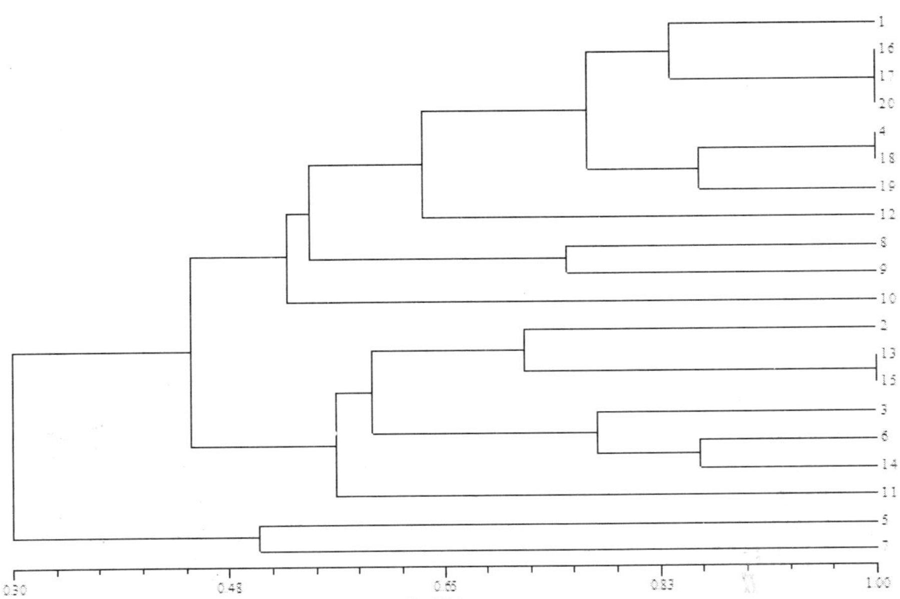

Figure 2 : Dendogram derived from banding patterns of RAPD analysis with primer OPAD 15.

(carbendazim tolerant) isolates were collected from Nauni, Solan and formed a small sub-group. RAPD primer OPAD -15 was efficient to group *Fusarium oxysporum* f.sp. *pisi* isolates on the basis of carbendazim tolerance and incubation period. Understanding the genetic structure of pathogen populations in present study may provide insights into the epidemiology and evolutionary potential of pathogen and could lead to improved strategies for controlling the disease.

References

Armstrong G M and Armstrong J K. (1974). Races of *Fusarium oxysporum* f.sp. *pisi*, causal agents of wilt of pea. *Phytopathology* 64:849-857.

Bauxton E W. (1956). Heterokaryosis and parasexual recombination in pathogenic strains of *Fusarium oxysporum*. *J Gen Microbiol* 15:133-139.

Chiocchetti A, Ghignone S, Minuto A, Gullino M L, Garibaldi A and Migheli Q. (1999). Identification of *Fusarium oxysporum* f. sp. *basilici* isolated from soil, basil seed and plants by RAPD analysis. *Plant Dis* 83:576-81.

Dice L R. (1945). Measures of the amount of ecological association between species. *Ecology* 26:297-302.

Falck R. (1907). *Waqchstumesetze, wachstumtakorch und temperature wertder holzensterenden*. *Mycelien* 1: 143-153.

Gupta S K and Upmanyu S. (2005). Management of *Sclerotium rolfsii*, causal agent of crown rot of French bean having decreased sensitivity to carbendazim. Paper presented in the second global conference "Plant health-global wealth" organized by the Indian Society of Mycology and Plant Pathology held at MPAUT, Udaipur w.e.f. Nov. 25-29, 2005.

Haglund W A and Kraft J M. (1970). *Fusarium oxysporum* f.sp. *pisi*, race 5. *Phytopathology* 60:1861-1880.

Kraft J M and Haglund W A. (1978). A reappraisal of the race classification of *Fusarium oxysporum* f.sp. *pisi*. *Phytopathology* 68:273-275.

Kraft J M. (1994). *Fusarium* wilt of peas (a review). *Agronomie* 14: 561-567.

Kumar A. (1993). *Studies on resistance to Fusarium wilt in some cultivars of peas (Pisum sativum L.)*. MSc. thesis, Dr. Y.S. Parmar University of Horticulture and Forestry, Solan (HP), India. 71p.

Linford M B. (1928). A *Fusarium* wilt of peas in Wisconsin, *Wis Agric Expt Sta Resi Bull* 85:44p.

Logrieco A, Bottalico A and Ricci V. (1990). Occurrence of *Fusarium* species and their mycotoxins in cereal grains from some Mediterranean countries. *Phytopathol Mediterr* 29: 81-89.

Murray MG and Thompson WF. (1980). Rapid isolation of high molecular weight plant DNA. *Nucleic Acid Research* 8:4321-4325.

Rohlf FJ. (1998). NTSYS-pc. Numerical Taxonomy and Multivariate Analysis System. Version 2.01, Setauket, New York.

Sneath PHA and Sokal RR. (1973). *Numerical Taxonomy.* Sneath PHA and Sokal RR. (ed) San Francisco, USA, Freeman and Company, pp

Snyder W C. (1933). A new vascular *Fusarium* disease of peas. *Science* 77:327.

Stefanelli D, Frontino M G and Lewis B G. (1996). *Fusarium oxysporum* f.sp. *pisi*; characterization of race 5 in Italy and possible presence of race 6. *Phytopathologia Mediterranea* 35:75-81.

Thind T S. (2008). Fungicide resistance: a perpetual challenge in disease control. *J Mycol Pl Pathol* 38(3): 407-418.

Verma Shalini and Dohroo N P. (2003). Morphological, cultural and pathogenic variability in *Fusarium oxysporum* isolates from pea. *Pl Dis Res* 18(1): 5-8.

Verma Shalini. (2001). Studies on *Fusarium* wilt of pea (*Pisum sativum* L.) Ph. D. thesis, Dr. Y.S. Parmar University of Horticulture and Forestry, Solan (HP), India. 155p.

Walker SL, Leath S, Hagler WM and Murphy JP. (2001). Variation among isolates of *Fusarium graminearum* associated with *Fusarium* head blight in North Carolina. *Plant Disease* 85:404-10.

Whitehead D S, Coddington A and Lewis B G. (1992). Classification of races by DNA polymorphism analysis and vegetative compatibility grouping in *Fusarium oxysporum* f.sp. *pisi*. *Physiol Mol Plant Path* 41:295-305.

Wicks T. (1974). Tolerance of the apple scab fungus to benzimidazole fungicides. *Pl Dis Reptr* 58: 886-889.

❑❑❑

Chapter 19

ISOLATION, BIOCHEMICAL AND MOLECULAR CHARACTERIZATION OF SALT-TOLERANT BACTERIA FROM RHIZOSPHERE OF *CYNODON DACTYLON*- A FACULTATIVE HALOPHYTE

U. CHAKRABORTY[1], B.N. CHAKRABORTY[2], S. ROY[1], P.L. DEY[2] AND A.P. CHAKRABORTY[2]

[1]Plant Biochemistry Laboratory, [2]Immuno-Phytopathology Laboratory, Department of Botany, University of North Bengal, Siliguri- 734013, West Bengal, India.

Introduction

Salinity is one of the most critical constraints which hampers agricultural productions many areas worldwide. However, utilization of salt-affected soils for agriculture is also indispensible to feed the burgeoning population of the world. Salinity may directly or indirectly inhibit cell division and the elongation growth state; therefore, leaves and stems of the affected plants appear stunted. On the other hand, plants suffering saline stress present alterations in their homeostasis, mainly because of a reduction in the osmotic potential and an inadequate ionic distribution, which causes a nutritional imbalance (Kafkafi and Bernstein, 1996; Singh and Chatrath , 2001).Primary salinization occurs naturally where the soil parent material is rich in soluble salts, or in the presence of a shallow saline groundwater lable. Secondary salinization occurs when significant amounts of water are provided by irrigation, with no adequate provision of drainage for the leaching and removal of salts, resulting in the soils becoming salty and unproductive. Salinity also increases the uptake of Na^+, which eventually results into decrease in uptake of Ca^{2+} and K^+ (Yildirim *et al.* 2006). Accumulation of excess Na^+ may cause metabolic disturbances in process where low Na^+ and high K^+ or Ca^{2+} are required for optimum function (Marschner 1995). Uptake and accumulation of Cl^- may disrupt photosynthetic function through the inhibition of nitrate reductase activity (Xu *et al.*2000). Although many new physical and chemical methods are being used along with traditional methods to bring salt- affected landmass under irrigation, not much success is viewed anywhere.

The biotic approach of plant-microbe interaction to overcome salinity problems has recently received considerable attention from many workers throughout the world. Certain soil bacteria can help plants to either avoid or partially overcome a variety of environmental stresses. Fewer reports are available regarding the siderophore and ACC deaminase enzyme production potential of PGPR under saline stress. Saline environment also harbour taxonomically diverse bacterial groups which exhibit modified physiological and structural characteristics under the prevailing saline conditions. To survive under extreme condition majority of these bacteria have developed diverse structure- function adaptive mechanisms including cell surface formation and changes in cellular phospholipid composition. They can also osmoregulate by synthesizing specific compatible organic osmolytes viz. glutamine, proline, glycine-betaine and a few of them accumulate inorganic solutes like Na^+, K^+. The regulation of high concentration of ions in the cytoplasm helps in maintaining turgor pressure, which is a pre requisite for growth under the conditions of elevated external osmotic pressure. The tolerance to broad salinity ranges and halophilic features allow the bacteria to inhabit any natural saline habitat. Keeping these in mind, in the present study emphasis was laid up on the isolation and characterization of salt tolerant rhizobacteria from the saline rhizosphere of *Cynodon dactylon* , a facultative halophyte and their further molecular characterization through RAPD profiles for better understanding the versatilities among the isolates.

Material and Methods

Plant material : *Cynodon dactylon,* known as facultative halophyte, naturally grown in the North Bengal University Campus was selected in order to isolate halo-tolerant bacterial isolates from its rhizosphere.

Collection of soil samples : Soil samples were collected from rhizosphere of *Cynodon* from two different locations.

Isolation of Bacterial strains : For the isolation, 1gm of soil was diluted with 25ml of sterile distilled water and kept for 30 mins in rotating conditions. Nutrient agar supplemented with 10% NaCl was used as a selective medium. One ml of soil suspension from aliquot dilutions (10^{-2} to 10^{-6}) was aseptically added. 15 ml of sterilized, melted and cooled medium was added and poured in each petriplate and were incubated at 37^{0}C for 24h. Following the appearance of colonies after some days, isolated colonies were picked up with sterilized loop, transferred on to fresh NA slants with 10% NaCl and the pure cultures so obtained were stored in refrigerator at 4^{0}C. Subsequent subculturing was also done in NaCl free nutrient agar and broth media for morphological, biochemical and molecular analyses.

Confirmation of the bacterial isolates as halotolerant : Mannitol salt agar (MSA) as described by Hitchens *et al.* (1995) was used to confirm the salt tolerance ability of the isolated bacteria. Mannitol Salt Agar is a nutritive medium due to its content of peptones and beef extract, which supply essential growth factors, such as nitrogen, carbon, sulfur and trace nutrients. The 7.5% concentration of sodium chloride resulted in the partial or complete inhibition of bacterial organisms other than halotolerant bacteria. Development of halo region due to the fermentation of mannitol, around the growth of microorganisms was a positive test for salt tolerant bacteria.

Morphological studies : Pure cultures of the isolates were streaked on NA plates with 10% NaCl separately for colony development. The individual colonies were examined for shape, size, structure of colonies and pigmentation. The Gram reactions of all the isolated isolates were recorded according to Buchanan and Gibbson (1974). Two loopful of the bacterial suspension was smeared on grease free clean micro slide, stained accordingly and observed under compound microscope. Gram positive/ negative reactions, shape of cells observed were recorded.

Biochemical studies : Biochemical tests viz, starch hydrolysis, catalase, indole production, urease, gelatinase, protease (Walsh *et al.*

1995), citrate, nitrate reduction, porphyrin reaction were carried out for biochemical confirmation of these isolates.

Phosphate solubilization : Phosphate solubilization by the isolated NaCl-tolerant bacteria was checked by allowing the bacteria to grow in selective medium i.e., Pikovskaya's agar (Pikovskaya, 1948). The appearance of transparent halo zone around the bacterial colony indicated the phosphate solubilizing activity of the bacteria.

Siderophore production : Production of siderophore was detected by standard method of Schwyn and Neiland(1987) using blue indicator chrome azurol S (CAS). The bacteria were spot inoculated at the center of the plate and incubated for 12-15 days. The change in the colour of the medium around the bacterial spot was an indication of siderophore production.

Chitinase production : Production of chitinase was detected by standard method of Hsu and Lockwood (1975). Colonies showing zones of clearance against the creamy background were regarded as chitinase-producing strains.

Protein content of bacterial isolates : The nine isolated bacteria were grown separately in Nutrient broth medium supplemented with 10% NaCl for 24 hrs. Then the cultures were centrifuged and bacterial pellets were suspended in 0.05M Na- phosphate buffer, pH-7.2 in order to sonicate the bacterial cells. After the sonication, bacterial protein was kept at -20 ^{0}C and was used for assay (Lowry *et al.* 1951) as well as SDS-PAGE analysis to compare the protein band patterns among the isolates.

Molecular variability among the isolates

Genomic DNA extraction

The broth cultures of bacterial isolates were centrifuged at 10,000 rpm at 28^{0}C for 5 mins and the pellets were collected by discarding the supernatant. The pellets were washed thrice with distilled water and resuspended in 0.5ml of lysis solution(100mM Tris Hcl, pH 7.5, 20mM EDTA, 250mM NaCl, 2% SDS, 1mg/ml lysozyme). To it 5 μl of RNase(50mg/ml) was added and incubated at 37^{0}C for 3 hrs. Then 10 μl proteinase K solution (20mg/ml) was added and it was allowed to incubate at 65^{0}C for 3min. The lysate was extracted with equal volume of tris water saturated phenol: chloroform: isoamylalcohol (25:24:1) and then centrifuged at 10,000 rpm for 5 min. The aqueous phase was collected in clean tube and 2 volume of chilled absolute ethanol was added to this aqueous phase. The mixture was centrifuged at 10,000 rpm for 5 mins at 4C, the pellet was air dried and finally dissolved in 40μl TE buffer and stored at 4C.

Agarose gel elctrophoresis : The quality of the genomic DNA, isolated from nine different isolates was checked on 0.8% agarose gel electrophoresis. The DNA from all isolates produced clear sharp bands, indicating good quality of DNA.

PCR amplification : All isolates were taken up for RAPD-PCR amplification. Genomic DNA was amplified by mixing the template DNA, with the polymerase reaction buffer, dNTP mix, primers and Taq polymerase. Polymerase chain reaction was performed in a total volume of 100ml, containing 78ml deionized water, 10ml 10X taq polymerase buffer, 1μl of 1U Taq polymerase enzyme, 6μl 2mM dNTPs, 1.5μl of 100mM RAPD primers and 1μl of template DNA. Three random decamers (OPA1, OPA4, AA1) were used to prepare the RAPD profiles of the isolates. PCR was programmed with an initial denaturing at 94°C for 5 min, followed by 30 cycles of denaturation at 94°C for 30sec, annealing at 59°C for 30 sec and extension at 70°C for 2 min and the final extension at 72°C for 7 min in a Primus 96 advanced gradient Thermocycler. After RAPD-PCR amplifications, all amplified DNA products were resolved by electrophoresis on agarose gel (2%) in TAE(1X) buffer, stained with ethidium bromide and photographed. After that all reproducible polymorphic bands were scored and analyzed following UPGMA cluster analysis protocol and computed in to similarity matrix using NTSYS computer program to prepare a dendrogram. The nucleotide sequences used for RAPD PCR are-

Seq Name	Primer Seq 5'-3'	Mer	TM	% GC
RAPD primers				
OPA-4	AATCGGGCTG	10	39.3	60%
OPA1	CAGGCCCTTC	10	38.2	70%

Results

Isolation and identification of bacterial strains : Out of nine strains, isolated from *Cynodon* rhizosphere, two of them (S3 and S5) were found to be a gram positive bacteria and others were identified as gram negative group. On the basis of morphological and biochemical observation, two isolates (S3 and S5) could be identified as *Staphylococcus* group of bacteria. These isolates produced yellowish zones on mannitol salt agar.

Morphological characterization : Based on the morphological characteristics, nine halo-tolerant isolates were studied in detail for colony, growth type, gram reaction and cell shape (Table 1 and Fig 1).

Table 1 : Cultural characteristics of different halo-tolerant isolates

Isolates	Gram reaction	Cell shape	Cell size(μm)	Margin of colony	Elevation	Colour of colony	Colony size(mm)	Type of growth	Color in MSA plate
S1	-ve	Rod with tapering end	1.14-1.5	Smooth	Convex	Yellow	0.2-0.8	Slow	White
S2	-ve	Rod	1.66-4	Smooth	Convex	White	0.8-1.5	Slow	White
S3	+ve	Coccus in chain	0.88-1.21	Smooth	Raised	White	0.6-1.8	Medium	White
S4	-ve	Short Rod	2-2.5	Smooth	Raised	Yellow red	0.1-0.8	Slow	White
S5	+ve	Coccus in chain	1.06-1.18	Undulated	Flat	White	0.3-1.0	Fast	White
N2	-ve	Rod	2.01-2.35	Smooth	Flat	Gray white	0.1-0.5	Slow	White
N3	-ve	Rod	2-3.6	Smooth	Raised	White	0.5-0.8	Medium	Gray white
N4	-ve	Rod	1.5-2.8	Undulated	Convex	Off white	0.1-1.9	Slow	White
N5	-ve	Short Rod	1.01-1.2	Smooth	Raised	Light red	0.2-1.3	Medium	Pale Red

Biochemical characterization : All the isolates produced catalase as well as reduced nitrate to nitrite, but showed negative results in urease tests. However, intensity of production of O_2 by catalase differed among the isolates. Only four isolates(S4,N3,N4 and N2) could hydrolyse starch, five isolates(S2,S5,N2,N3 and N5)

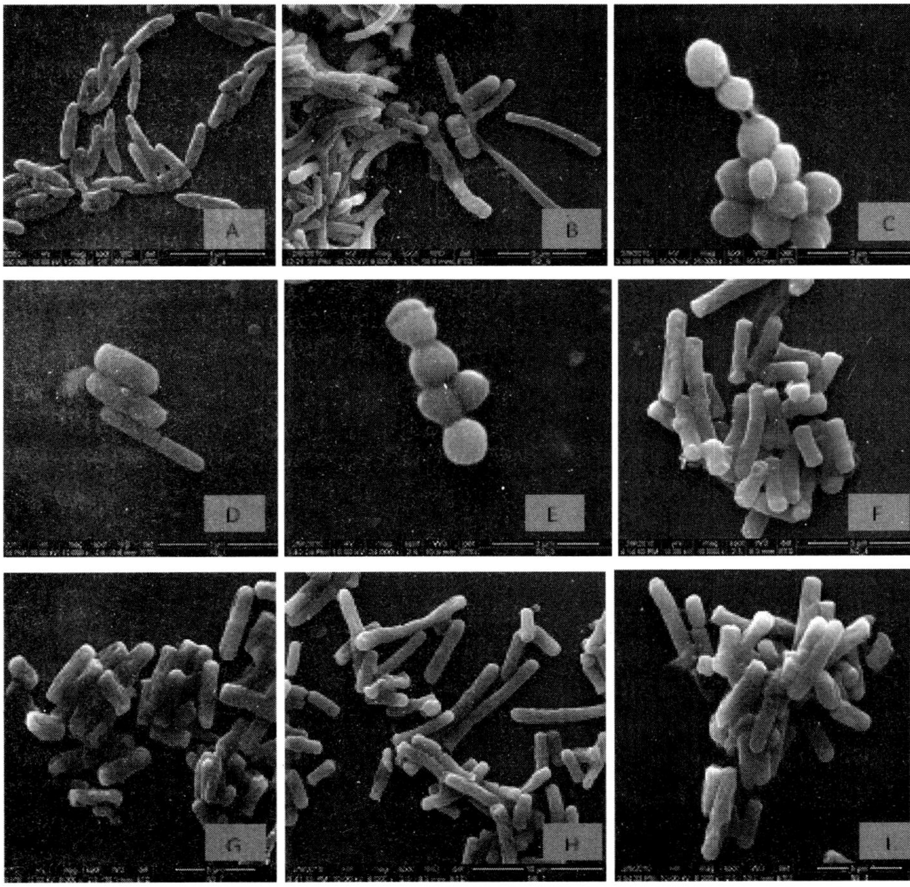

Figure 1 : Scanning electron microscopic view of salt-tolerant bacteria; A-S1,B-S2,C-S3,D-S4,E-S5, F-N2, G-N3, H-N4 and I-N5

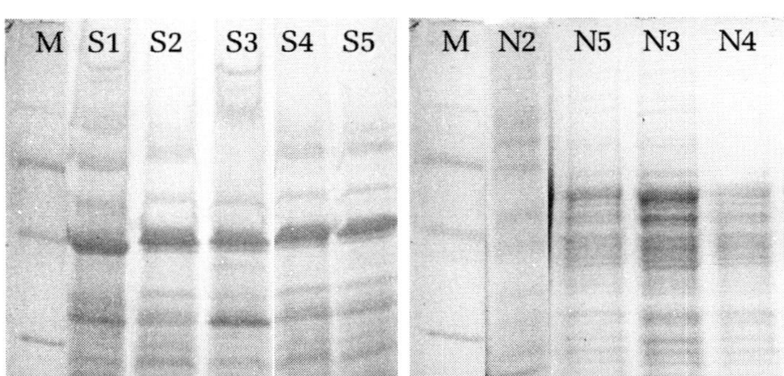

Figure 2 (A&B) : SDS- PAGE analysis of proteins of isolated nine salt-tolerant bacterial isolates

liquefied gelatin while all had the ability to produce indole. Six isolates showed the potentiality in protease (S3, S4, S5, N2, N3 and N4) as well as citrate tests (S2, S4, S5, N3, N4 and N5). Seven out of nine isolates gave positive responses in porphyrin test but strong reaction was noticed in S4, N3 and N5 isolates. S1 and S5 isolates showed negative responses in porphyrin tests (Table 2).

Table 2 : Biochemical responses of different halo-tolerant isolates

Biochemical tests performed	Bacterial isolates								
	S1	S2	S3	S4	S5	N2	N3	N4	N5
Catalase	++	+++	+	+	+++	+	+	+	+++
Nitrate Reduction	+	+	+	+	+	+	+	+	+
Urease	-	-	-	-	-	-	-	-	-
Starch hydrolysis	-	-	-	+	-	+	+	+	-
Gelatin liquefication	-	+	-	-	+	+	+	-	+
Indole production	+	+	+	+	+	+	+	+	+
Protease	-	-	+	+	+	+	+	+	-
Citrate	-	+	-	+	+	-	+	+	+
Porphyrin	-	-	-	++	-	-	++	-	++
Phosphate solubilization	-	-	-	+	-	-	-	-	-
Chitinase	-	+	-	+	-	+	+	+	-
Siderophore production	-	-	-	+	-	+	-	+	-

SDS- PAGE analysis of halo-tolerant isolates

Analysis of SDS-PAGE of soluble proteins, made from the isolates showed distinct protein patterns, with molecular masses ranging from 12-106 KDa (Fig 2).

RAPD analysis : The genetic relatedness among nine salt-tolerant bacterial isolates were analysed by two random primers OPA-1 and OPA-4 to generate reproducible polymorphisms. All amplified products with the primers had shown polymorphic and distinguishable banding patterns which indicate the genetic diversity of isolates. RAPD profiles showed that primer OPA-4 scored highest bands (Fig 3). Relationships among the isolates was evaluated by cluster analysis of the data based on the similarity matrix. The dendogram was generated by unweighted pair- group methods with arithmetic mean (UPGMA) using BIO profil ID image software for each primer and NTSYSpc software (Fig 4). Similarity co-efficient ranged from 0.65-1.00. Based on the results obtained all nine isolates can be grouped into 4 main clusters. One cluster represents S1 isolate, 2nd one represents S2,S4,S5,S3 and N5 isolates. Again 2nd cluster is also sub grouped into four, first subgroup with S2&S4 isolates, second, third and fourth one represent S5, S3 and N5 halo-tolerant isolates. Third cluster is with 3 isolates, which is further subdivided into two, first with N2 & N3 and second one represents N4 isolate.

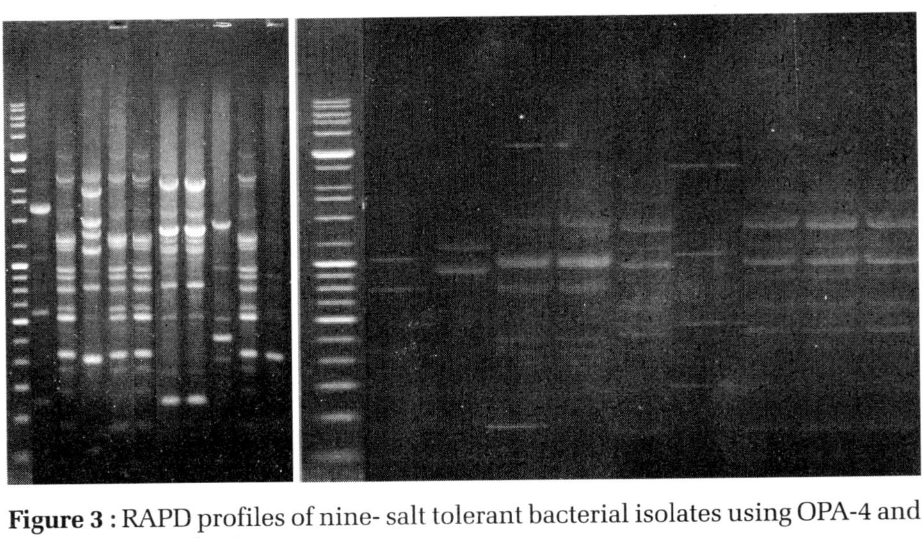

Figure 3 : RAPD profiles of nine- salt tolerant bacterial isolates using OPA-4 and OPA-1 universal primers.

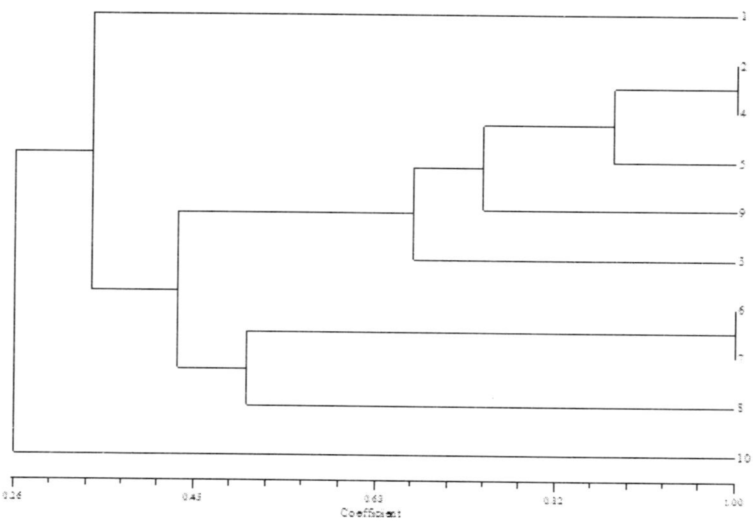

Figure 4 : Dendogram showing relationships among the isolates.

Discussion

Saline conditions are known to suppress the growth of plants, causing a diminished yield. An important factor to be considered when screening new isolates is their activity in the range of environments in which they would be expected to be used. Taking this into account, we isolated native strains from saline soils of *Cynodon dactylon* and screened for salt-tolerant bacteria. Nine bacterial isolates (S1, S2, S3, S4, S5, N2, N3, N4 and N5) were isolated and these isolates were further confirmed in mannitol salt agar for their salt-tolerance characteristics. From the morphological and biochemical point of view, two isolates (S3 and S5) could be identified as *Staphylococcus* group of bacteria. The siderophore produced by bacteria may contribute to an increased mobility of Fe in the soil and rhizosphere, making it more available for the plant. All the isolates were screened for siderophore production *in vitro*, but only three isolates (S4,N2 and N4) showed positive responses. It has been reported that *Azospirillum brasilense* increased the iron absorption and translocation by sorghum (Barton LL *et al.* 1986). None of the isolates had the ability to produce chitinase *in vitro*. Another essential macronutrient is phosphorus, which is frequently immobilized and unavailable for plants. Some PGPR are able to solubilize phosphate, such as *Azotobacter chroococcum* (Kumar and Narula. 1999), *Bacillus sp.* and *Pseudomonas putida* (Cattelan *et al.* 1999). Only one isolate (S4) solubilized phosphate.

SDS- PAGE analysis of proteins showed different band patterns among the isolates as well as the difference in the intensity of bands. The genetic relatedness among nine salt-tolerant bacterial isolates were analysed. All amplified products with the primers had shown polymorphic and distinguishable banding patterns which indicate the genetic diversity of isolates. Relationships among the isolates was evaluated by cluster analysis of the data based on the similarity matrix. Similarity co-efficient ranged from 0.65-1.00. Based on the results obtained all nine isolates can be grouped into 4 main clusters. One cluster represents S1 isolate, 2nd one represents S2,S4,S5,S3 and N5 isolates. Third cluster is with 3 isolates (N2, N3 and N4).

From the results it may be concluded that nine native strains, isolated from soil of facultative halophyte- *Cynodon dactylon* were characterized biochemically for salt- tolerance. On the basis of biochemical characterization, two isolates (S3 and S5) may belong to *Staphylococcus* group. Further molecular characterization of these isolates showed the relationships among the isolates. This is the preliminary work, but further studies are being carried out to use S4 isolate out of the nine isolates since it showed both salt-tolerance and plant growth promoting activities (produce siderophore, solubilize phosphate). So, application of S4 in plants in saline soils or salt stressed plants will be helpful to promote plant growth in stressed conditions.

Acknowledgement

Financial support received from CSIR, New Delhi is gratefully acknowledged.

References

Barton LL, Johnson GV and Miller SO. (1986). The effect of *Azospirillum silense* on iron absorption and translocation by sorghum. *J Plant Nutr.* 9: 557–565.

Buchanan RE and Gibbson NE. (1974). Bergeys Manual of Determinative Bacteriology. 8th ed. Williams and Wilkins Co. Baltimore.

Cattelan A, Hartel P and Fuhrman J. (1999). Screening for plant growth-promoting rhizobacteria promote early soybean growth. *Soil Sci Soc Am J.* 63: 1670–1680.

Hitchens, Tran and McCarron. (1995). FDA bacteriological analytical manual, 8th ed. AOAC International, Gaithersburg, Md.

Hsu SC and Lockwood JL. (1975). Powdered chitin agar as a selective medium for enumeration of actinomycetes in water and soil. *Appl. Microbiol.* 29: 422-426.

Kafkafi U and Bernstein N. (1996). Root growth under salinity stress. In: Waisel Y, Eshel A, Kafkafi U (eds) Plant roots: the hidden half. New York: Marcel Decker, pp 435–451.

Kumar V and Narula N. (1999). Solubilization of inorganic phosphates and growth emergence of wheat as affected by *Azotobacter chroococcum* mutants. *Biol Fertil Soils.* 28:301–305.

Lowry OH, Rosebrough NJ, Fair AL and Randall RJ. (1995). Protein measurements with Folin Phenol reagent. *J. Biol. Chem.* 193:265-275.

Marschner H. (1995). Mineral nutrition of higher plants. London: Academic Press. P 889.

Marulanda-Aguirre A, Azcon R, Ruiz-Lozano JM and Aroca R.(2008). Differential effects of a *Bacillus megaterium* strain on *Lactuca sativa* plant growth depending on the origin of the arbuscular mycorrhizal fungus coinoculated: physiologic and biochemical traits. *J Plant Growth Regulation.* 27:10-18.

Pikovskaya RI. (1948). Mobilization of phosphorous in soil connection with the vital activity of some microbial species. *Microbiologiya.* 17: 362-370.

Schwyn B. and Neiland JB. (1987). Universal chemical assay for the detection and determination of siderophores. *Anal Biochem.* 160: 47-56.

Singh KN and Chatrath R. (2001). Salinity tolerance. In: Reynolds MP, Ortiz-Monasterio JI, McNab A (eds) Application of physiology in wheat breeding. M_xico, D.F: International Maize and Wheat Improvement Center (CIMMYT), pp 101–110.

Walsh GA, Murphy RA, Killen GF and Headon Power DR RF. (1995). Technical note: detection and quantification of supplemental fungal â-glucanase activity in animal feed. *J Animal Sci.* 73:1074–1076.

Xu G, Magen H, Tarchirky J and Kafkafy U. (2000). Advances in chloride nutrition in plants. *Adv Agron.* 68: 97-150.

Yildirim E, Taylor AG and Spittler TD. (2006). Ameliorative effects of biological treatments on growth of squash plants under salt stress. *Scientia Hort.* 111. 1-6.

❑❑❑

Chapter 20

BIOCHEMICAL AND MOLECULAR CHARACTERIZATION OF *STREPTOMYCES* SP. ISOLATED FROM AGRICULTURAL FIELD OF NORTH BENGAL

P.L.DEY, P.RAY, U. CHAKRABORTY AND B.N.CHAKRABORTY

Immuno-Phytopathology Laboratory, Department of Botany,
University of North Bengal, Siliguri-734013, West Bengal

Introduction

Actinomycetes are gram positive bacteria having mycelial growth. They produce branching mycelia of two kinds; aerial mycelia and substrate mycelia. Actinomycetes are filamentous, branching bacteria with a fungal type of morphology.They are part of the microbial flora of most natural substrates (Moustafa *et. al*, 1963) and mainly found in the rhizosphere of plants in association with other microorganisms like rhizobacteria and fungi. They utilize humic acid and other organic matter in soil. In their natural habitat, such as forests, the actinomycetes interact in various ways with the higher plants (Lo *et. al*, 2002). Among actinomycetes, the streptomycetes are dominant. These organisms are part of PGPM or plant growth promoting microorganisms. One of the plant growth promoting factors present in PGPM is the ability to solubilise phosphate present in soil. Phosphate is an important part of plant system. In its absence plant growth is retarded. Phosphate solubilizing microbes can play an important role in plant nutrition through an increase in phosphate uptake by plants (Antoun 1998). Streptomycetes having the phosphate solubilizing ability can help the plant to get more phosphate from the soil thereby increasing plant growth. Since a large number of antibiotics are produced by streptomycetes (Yuan *et. al*, 1995), they also act as an antagonist that can prevent pathogens causing disease in plants (Crawford *et. al,* 1993).

The present study aims at identifying streptomycetes isolates from the agricultural fields of North Bengal having plant growth promoting characteristics like Phosphate solubilizing activity as well as antagonistic effect against plant pathogens.

Materials and Methods

Collection of soil samples : Soil samples were collected from various locations of Jalpaiguri and Darjeeling district. Soil samples were obtained from a depth of 6-10 cm in the rhizosphere of agricultural crops. The soil samples were allowed to air dry at room temperature for a week. Soil samples were collected from the rhizosphere soil of potato plants.

Isolation : The standard serial dilution plate technique by Warcup (1955) was used for isolation of streptomycetes. 5 grams of soil sample was suspended in 25 ml sterile distilled water. The suspension was stirred for 3 hour in magnetic stirrer and left for ½ hour so that soil particles come down. The supernatant was used as stock solution which was diluted upto 10^{-6} level by serial dilution method. 1 ml of the dilutions were spread on Starch Casein Nitrate agar (SCN) medium, the plates were incubated at 35^{0}C for 7 days.

Biochemical tests : Biochemical tests which are confirmatory for streptomycetes were executed to determine whether the isolates have the ability to hydrolyze starch, production of Catalase and production of indole ring. .

Starch hydrolysis : The isolates were streaked on sterilized starch agar plate (NA + 0.1% soluble starch) and incubated for five days at 37° C. The plates were flooded with Lugol's iodine solution. The clear zone underneath and around the growth indicates the starch hydrolysis

Indole test : 10ml of Davis Mingoli's broth supplemented with 0.1% tryptophan was inoculated with the isolate and incubated anaerobically at 37° C for 7 days. The culture were layered carefully with 2 ml of Ehrlich- Bobme (P-dimethylaminobenzaldehyde 10g, concentrated HCL 100ml) reagent on the surface, allowed to stand for a few minutes and observed for the formation of a ring at the medium reagent interface indicating the production of indole.

Catalase : Culture (24 hour old) was flooded with 0.5 ml 10% H_2O_2 solution and gas bubbles production indicated the positive reaction.

Screening for phosphate solubilizing activity : Streptomycetes isolates were screened in Pikovskaya medium for phosphate solubilization activity (Pikovskaya, 1948). Isolates were inoculated in Pikovskaya media and incubated for 7 days. A halo zone around the growth indicates phosphate solubilization activity.

In vitro antagonistic effect on fungal pathogen :Streptomycetes isolates were tested for antagonistic effect against fungal pathogens of *Fusarium graminarium, F. solani* and *Sclerotium rolfsii* by dual culture method (Skidmore and Dickinson (1976). Streptomycetes isolates were placed in both PDA and SCN plates in a circle and the set up was incubated for 8 days. Blocks cut off from 8 day old fungal pathogen culture were transferred to the center of Petri dishes inoculated with streptomycetes isolates. Inhibitions of the radial growth of fungal pathogen by the isolates confirm antagonism.

Molecular variability among the isolates

Genomic DNA extraction : The broth cultures of isolates were centrifuged at 10,000 rpm at 28°C for 5 mins and the pellets were collected by discarding the supernatant. The pellets were washed thrice with distilled water and resuspended in 0.5ml of lysis solution(100mM Tris Hcl, pH 7.5, 20mM EDTA, 250mM NaCl, 2% SDS, 1mg/ml lysozyme). To it 5 μl of RNase(50mg/ml) was added and incubated at 37°C for 3 hrs. Then 10 μl proteinase K solution (20mg/ml) was added and it was allowed to incubate at 65°C for 3min. The lysate was extracted with equal volume of tris water saturated phenol: chloroform: isoamylalcohol (25:24:1) and then centrifuged at 10,000 rpm for 5 min. The aqueous phase was collected in clean tube and 2 volume of chilled absolute ethanol was added to this aqueous phase. The mixture was centrifuged at 10,000 rpm for 5 mins at 4°C, the pellet was air dried and finally dissolved in 40μl TE buffer and stored at 4°C.

Quantification of genomic DNA by gel elctrophoresis : The quantity and quality of the genomic DNA, isolated from nine different isolates was checked on 0.8% agarose gel electrophoresis. The DNA from all isolates produced clear sharp bands, indicating good quality of DNA.

PCR amplification : All isolates were taken up for PCR amplification. Genomic DNA was amplified by mixing the template DNA, with the polymerase reaction buffer, dNTP mix, primers and Taq polymerase. Polymerase chain reaction was performed in a total volume of 100μl, containing 78ml deionized water, 10μl 10X taq polymerase buffer, 1μl of 1U Taq polymerase enzyme, 6μl 2mM dNTPs, 1.5μl of 100mM primer pairs each and 1μl of template DNA. Genus specific primer pairs R1 and F5 were used for amplification. PCR was programmed with an initial denaturing at 94°C for 5 min, followed by 30 cycles of denaturation at 94°C for 30sec, annealing at 61°C for 30 sec and extension at 70°C for 2 min and the final extension at 72°C for 7 min in a Primus 96 advanced gradient Thermocycler. After PCR amplifications, all amplified DNA products were resolved by electrophoresis on agarose gel (1%) in TAE(1X) buffer, stained with ethidium bromide and photographed.

Results and Discussion

Isolation of streptomycetes : A total number of 18 isolates were isolated from starch casein nitrate agar medium. The GPS location of sampling area were recorded with GPS Garmin tool. Many of the isolates show typical morphology of streptomycetes group. The colony colour varies from white, pink, grey, red. Some of the isolates produced soluble pigments which diffuse in the media. (Table 1).

Biochemical tests : All of the isolates showed positive result for biochemical tests which confirmed that the isolates are streptomycetes. (Table 2).

Table 1 : GPS location and colony characteristics of Streptomycetes isolated from potato rhizosphere

Isolates	GPS Location	Colony Properties
Streptomyces sp (RHS-PO-11)	N 26°44'54.08" E 88°48'14.53"	Reddish white
Streptomyces sp (ARHS-PO-12)	N 26°44'54.88" E 88°48'14.04"	Reddish
Streptomyces sp (ARHS-PO-13)	N 26°44'53.18" E 88°48'13.24"	Whitish
Streptomyces sp (ARHS-PO-14)	N 26°44'54.89" E 88°48'14.34"	Pinkish white
Streptomyces sp (ARHS-PO-15)	N 26°44'57.23" E 88°48'15.24"	Reddish white
Streptomyces sp (ARHS-PO-16)	N26°47'09.44" N88°22'06.53"	Pinkish
Streptomyces sp (ARHS-PO-17)	N 26°44'54.08" E 88°48'14.53"	Reddish
Streptomyces sp (ARHS-PO-18)	N 26°44'54.88" E 88°48'14.04"	Grayish white
Streptomyces sp (ARHS-PO-20)	N 26°44'53.18" E 88°48'13.24"	Grayish
Streptomyces sp (ARHS-PO-21)	N 26°44'54.89" E 88°48'14.34"	Grayish white
Streptomyces sp (ARHS-PO-22)	N 26°44'57.23" E 88°48'15.24"	Whitish
Streptomyces sp (ARHS-PO-23)	N26°47'09.44" N88°22'06.53"	Pinkish
Streptomyces sp (ARHS-PO-24)	N 26°44'54.08" E 88°48'14.53"	Grayish
Streptomyces sp (ARHS-PO-25)	N 26°44'54.88" E 88°48'14.04"	Greywith red periphery
Streptomyces sp (ARHS-PO-26)	N 26°44'53.18" E 88°48'13.24"	Pinkish
Streptomyces sp (ARHS-PO-27)	N 26°44'54.89" E 88°48'14.34"	Grayish
Streptomyces sp (ARHS-PO-28)	N 26°44'57.23" E 88°48'15.24"	Whitish
Streptomyces sp (ARHS-PO-29)	N26°47'09.44" E 88°48'14.04"	Whitish

Phosphate solubilization activity : 15 isolates out of 18 showed phosphate solubilization activity by the production of clear zone around the growth in Pikovskaya medium. (Table 2). Goodfellow *et al.* (1987) investigated the use of rapid biochemical tests for streptomycete taxonomy by using a fluorogenic probe which was conjugated to the substrate and fluoresced after hydrolysis by the target enzyme. Streptomycete strains from 17 taxonomic clusters as defined by Williams *et al.* (1983) were investigated for 14 different enzyme activities. A discontinuous distribution of enzymes was detected in all of the test strains, which raises questions as to the usefulness of this method as a taxonomic tool for the genus *Streptomyces*.

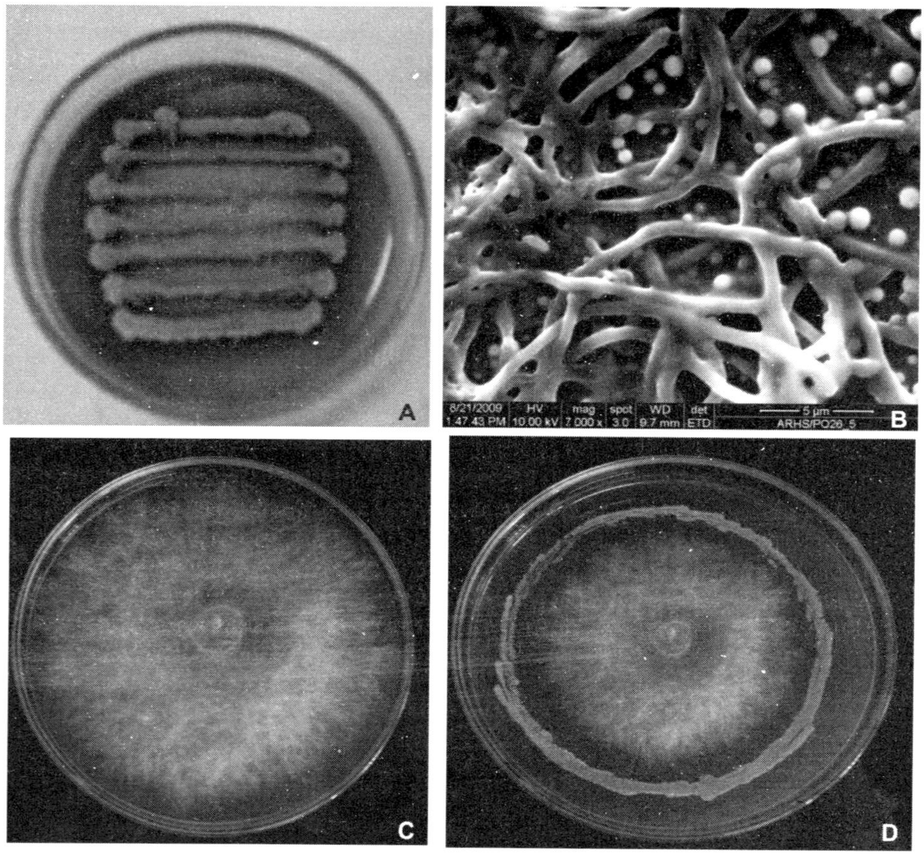

Figure 1 : *Streptomyces griseus* in SCN medium (a), SEM of *Streptomyces griseus* (B), antifungal activity of *Streptomyces griseus* against *Sclerotium rolfsii* ; Control (C) antifungal activity (D)

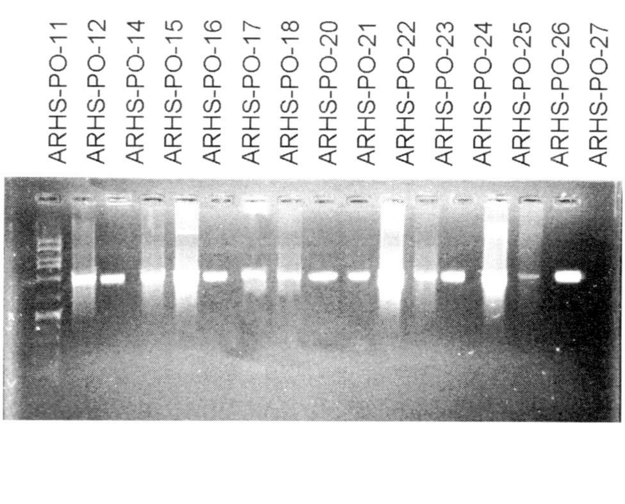

Figure 2 : 1% Agarose gel electrophoresis of PCR amplified products of 16S rDNA region of *Streptomyces sp.*

Table 2 : Biochemical tests of *Streptomyces* sp.

***Streptomyces* sp. Isolates**	**Catelase test**	**Indole test**	**Starch hydrolysis**	**Phosphate solubilizing activity**
ARHS-PO-11	+	+	+	+
ARHS-PO-12	+	+		+
ARHS-PO-13	+	+	+	-
ARHS-PO-14	+	+	+	+
ARHS-PO-15	+	+	+	-
ARHS-PO-16	+	+	+	+
ARHS-PO-17	+	+	+	+
ARHS-PO-18	+	+	+	+
ARHS-PO-20	+	+	+	+
ARHS-PO-21	+	+	+	+
ARHS-PO-22	+	+	+	+
ARHS-PO-23	+	+	+	+
ARHS-PO-24	+	+	+	+
ARHS-PO-25	+	+	+	+
ARHS-PO-26	+	+	+	+
ARHS-PO-27	+	+	+	-
ARHS-PO-28	+	+	+	+
ARHS-PO-29	+	+	+	+

In vitro antagonistic effect on fungal pathogen : Isolates were tested for *in vitro* antagonistic effect against selected plant pathogenic fungi.among the 18 isolates 3 isolates, ARHS-PO-12, ARHS-PO-26, and ARHS-PO-27 showed antagonistic activity. Whereas ARHS-PO-12 and ARHS-PO-27 showed activity against *F. solani*, ARHS-PO-26 was effective to control *S. rolfsii.* (Table 3).

PCR amplification : Genomic DNA of 15 phosphate solubilizing streptomycetes isolates were prepared and their 16S rDNA was amplified using genus specific primers (R1 and F5) . PCR products were subjected to horizontal electrophoresis. The primes were used to amplify nearly full length 16S rDNA sequences and yielded 1.5kb product. These results are also in accordance with several workers who observed the amplified rDNA fragment of approximately 1.5 kbp by PCR (Fig 2). These rDNA are highly stable and exhibit a mosaic of

Table 3: Result of *in vitro* antagonistic effect of actinomycetes on fungal pathogens:

Streptomycetes isolate	*Fusarium graminarium* Inhibition (%)	*F.solani* Inhibition (%)	*Sclerotium rolfsii* Inhibition (%)
ARHS-PO-11	74.4	66.7	67.8
ARHS-PO-12	75.6	85.6	56.7
ARHS-PO-13	77.8	67.8	66.7
ARHS-PO-14	75.6	56.7	57.8
ARHS-PO-15	75.6	66.7	56.7
ARHS-PO-16	74.4	57.8	66.7
ARHS-PO-17	85.6	56.7	67.8
ARHS-PO-18	75.6	66.7	60.0
ARHS-PO-20	77.8	67.8	60.0
ARHS-PO-21	67.8	60.0	62.2
ARHS-PO-22	56.7	60.0	60.0
ARHS-PO-23	66.7	62.2	62.2
ARHS-PO-24	57.8	60.0	67.8
ARHS-PO-25	56.7	62.2	60.0
ARHS-PO-26	87.6	85.6	87.8
ARHS-PO-27	67.8	85.6	67.8
ARHS-PO-28	56.7	67.8	56.7
ARHS-PO-29	66.7	56.7	66.7

conserved and diverse regions within the genome (Hibbett, 1992). They also occur in multiple copies with up to 200 copies per haploid genome (Bruns, 1991; Yao, 1992) arranged in tandem repeats with each repeat consisting of the 18S small subunit (SSU), the 5.8S, and the 28S large subunit (LSU) genes. Internal transcribed spacer (ITS) regions have been used successfully to generate specific primers capable of differentiating closely related fungal species (Bryan, 1995). In the broader context, taxon-selective amplification of ITS regions is likely to become a common approach in molecular identification strategies with BLAST method which showed significant homology within NCBI database (Fig 3). Sequence analysis of rRNA genes has been applied to streptomycete taxonomy to investigate relationships at the genus, species and strain level. The use of 16S rRNA is reviewed by Stackebrandt *et al.* (1992), highlighting the importance of the region

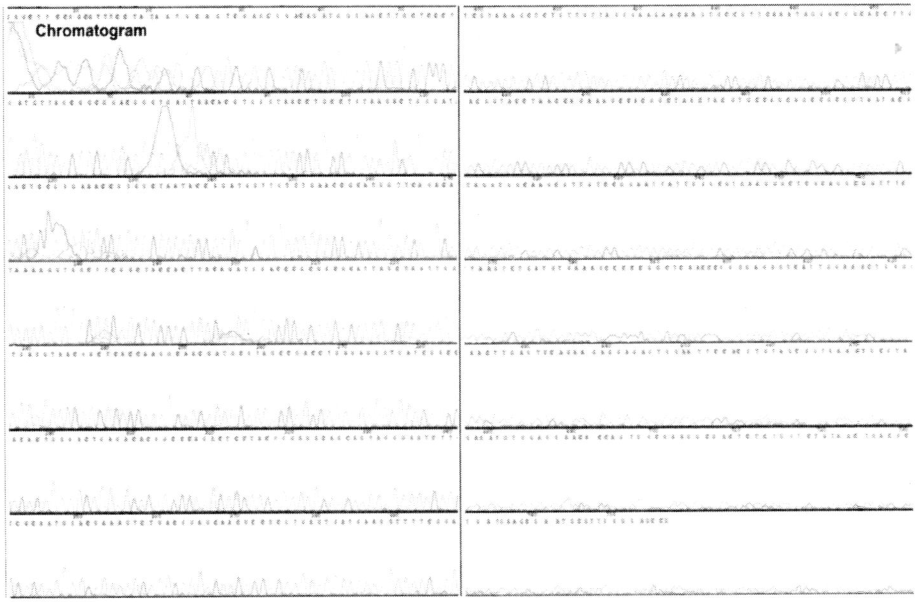

Partial sequence of 16S rDNA region

GGGCTTCCGGCGTTTCCTATAATGCAGTCGAGCGGACAGATGGGAGCTTGCTCCCTGATGTTAGCGGCGGAC
GGGTGAGTAACACGTGGGTAACCTGCCTGTAAGACTGGGATAACTCCGGGAAACCGGGGCTAATACCGGATG
GTTGTTTGAACCGCATGGTTCAGACATAAAAGGTGGCTTCGGCTACCACTTACAGATGGACCCGCGGCGCAT
TAGCTAGTTGGTGAGGTAACGGCTCACCAAGGCAACGATGCGTAGCCGACCTGAGAGGGTGATCGGCCACAC
TGGGACTGAGACACGGCCCAGACTCCTACGGGAGGCAGCAGTAGGGAATCTTCCGCAATGGACGAAAGTCTG
ACGGAGCAACGCCGCGTGAGTGATGAAGGTATATTGCACAATGGGCGAAAGCCTGATGCAGCGACGCCGCGT
GAGGGATGACGGCCTTCGGGTTGTAAACCTCTTTCAGCAGGGAAGAAGCGAAAGTGACGGTACCTGCAGAAG
AAGCGCCGGCTAACTACGTGCCAGCAGCCGCGGTAATACGTAGGGCGCAAGCGTTGTCCGGAATTATTGGGC
GTAAAGAGCTCGTAGGCGGCTTGTCACGTCGGTTGTGAAAGCCCGGGGCTTAACCCCGGGTCTGCAGTCGAT
ACGGGCAGGCTAGAGTTCGGTAGGGGAGATCGGAATTCCTGGTGTAGCGGTGAAATGCGCAGATATCAGGAG

Sequences producing significant alignments:

Accession	Description	Max score	Total score	Query coverage	E value	Max ident	Links
GU213508.1	Streptomyces sp. HB116 16S ribosomal RNA gene, partial sequence	2084	2084	100%	0.0	100%	
GU213499.1	Streptomyces sp. HB094 16S ribosomal RNA gene, partial sequence	2084	2084	100%	0.0	100%	
GQ924533.1	Streptomyces sp. ACT-0093 16S ribosomal RNA gene, partial sequence	2084	2084	100%	0.0	100%	
AM913957.1	Streptomyces sp. L108 partial 16S rRNA gene, isolate L108	2084	2084	100%	0.0	100%	
AM913955.1	Streptomyces sp. L105 partial 16S rRNA gene, isolate L105	2084	2084	100%	0.0	100%	
AM913869.1	Streptomyces sp. L148 partial 16S rRNA gene, isolate L148	2084	2084	100%	0.0	100%	
EU443837.1	Streptomyces sp. S240 16S ribosomal RNA gene, partial sequence	2084	2084	100%	0.0	100%	
EU263063.1	Streptomyces sp. MP47-91 16S ribosomal RNA gene, partial sequence	2084	2084	100%	0.0	100%	
EU158342.1	Streptomyces sp. HBUM 172143 16S ribosomal RNA gene, partial sequenc	2084	2084	100%	0.0	100%	
EF571003.1	Streptomyces flavofungini 16S ribosomal RNA gene, partial sequence	2084	2084	100%	0.0	100%	
EF571002.1	Streptomyces griseus strain 45H 16S ribosomal RNA gene, partial sequenc	[illegible]	[illegible]	[illegible]	[illegible]	[illegible]	
AF429294.1	Streptomyces sp. YTT E-99-1330 (A63) 16S ribosomal RNA gene, partial	2084	2084	100%	0.0	100%	
AF429393.1	Streptomyces sp. YTT E-99-1329 (A52) 16S ribosomal RNA gene, partial	2084	2084	100%	0.0	100%	
AJ844251.1	Actinomycetales bacterium N02 16S ribosomal RNA gene, partial sequenc	2084	2084	100%	0.0	100%	
NR_025870.1	Streptomyces sampsonii strain ATCC 25495 16S ribosomal RNA, complete	2084	2084	100%	0.0	100%	
EU263062.1	Streptomyces sp. MP47-06 16S ribosomal RNA gene, partial sequence	2078	2078	100%	0.0	99%	
GQ924642.1	Streptomyces champavatii strain NRRL B-5682 16S ribosomal RNA gene, p	2078	2078	100%	0.0	99%	
EU082956.1	Streptomyces sp. MEDD-5709A 16S ribosomal RNA gene, complete sequer	2074	2074	100%	0.0	99%	
GU213495.1	Streptomyces sp. HB105 16S ribosomal RNA gene, partial sequence	2073	2073	99%	0.0	100%	
AM412559.1	Streptomyces sp. 45zhy partial 16S rRNA gene, strain 45zhy	2073	2073	100%	0.0	99%	
GQ863925.1	Streptomyces sp. HB208 16S ribosomal RNA gene, partial sequence	2071	2071	99%	0.0	99%	
GU213501.1	Streptomyces sp. HB118 16S ribosomal RNA gene, partial sequence	2071	2071	99%	0.0	100%	
AM912954.1	Streptomyces sp. L104 partial 16S rRNA gene, isolate L104	2071	2071	99%	0.0	100%	
EU606725.1	Streptomyces albidoflavus strain TBG-S13A5 16S ribosomal RNA gene, pa	2069	2069	99%	0.0	100%	
AM912960.1	Streptomyces sp. L112 partial 16S rRNA gene, isolate L112	2069	2069	99%	0.0	100%	
EU158343.1	Streptomyces sp. HBUM 172144 16S ribosomal RNA gene, partial sequenc	2069	2069	100%	0.0	99%	
EU529871.1	Uncultured bacterium clone nbt98a02 16S ribosomal RNA gene, partial sec	2067	2067	100%	0.0	99%	

Figure 3 : Chromatogram and BLASTn result of 16S rDNA region of *Streptomyces gresius*.

selected for comparison. Relationships obtained differed according to the variable region (*a*, *b* and *c*) that was studied. Kataoka *et al.* (1997) conducted a comprehensive study of the *c* region from 89 streptomycete type strains that represented several of the major clusters as defined by Williams *et al.* (1983). The study verified that these variable regions can be used to resolve inter- and intraspecies relationships within the streptomycetes, despite being too variable for determining generic relationships. Of the 89 strains investigated, 57 variants were detected and 42 strains had unique sequences. Since that publication, the same group has subsequently sequenced the *c* regions from 485 streptomycete strains and deposited them in GenBank. This is the largest publically available set of streptomycete 16S rDNA sequence data. Comparison of the *c* regions from representatives of the major cluster groups defined by Williams *et al.* (1983) supports the taxonomic status of the phenotypic groups, the only strains with identical *c*regions were *Streptomyces olivaceoviridis* and *Streptomyces griseoruber* representatives from clusters 20 and 21, respectively. Analysis of the *Streptomyces albido flavus* group, which had previously been shown to comprise three species groups and over 60 strains (Williams *et al.*, 1989), was resolved into six groups using sequence comparisons of the *c* region (Kataoka *et al.*, 1997).

Acknowledgement

Financial support from ICAR, as a part of network project entitled Application of Microorganisms in Agriculture and Allied Sectors (AMAAS) of National Bureau of Agriculturally Important Microorganisms (NBAIM), is gratefully acknowledged.

References

Antoun, H., Beauchamp, C. J., Goussard, N., Chabot, R. & Lalande, R. (1998). Potential of Rhizobium and Bradyrhizobium species as growth promoting rhizobacteria on non-legumes: effect on radishes (*Raphanus sativus* L.). *Plant Soil* 204 : 57–67.

Crawford ,D.L. Lynch, J.M. Whipps ,J.M. and Ousley ,M. A.(1993). Isolation and characterization of Actinomycetes antagonists of a fungal pathogen. *Appl.Environ.Microbiol.* 59 (11) : 3899-3905.

Goodfellow, M., Lacey, J. & Todd, C. (1987). Numerical classification of thermophilic streptomycetes. *J Gen Microbiol* 133 : 3135-3149.

Kataoka, M., Ueda, K., Kudo, T., Seki, T. & Yoshida, T. (1997). Application of the variable region in 16S rDNA to create an index for rapid species identification in the genus *Streptomyces*. *FEMS Microbiol Lett* : 151: 249-255.

Lo, C. W, Lai ,N.S, Cheah H.Y, Wong, N.K.I and Ho ,C.C. (2002). Actinomycetes isolated from soil samples from the crocker range sabah. *ASEAN Review of Biodiversity and Environmental Conservation* (ARBEC).

Moustafa A, El-nakeeb and Hubert, A Lechevalier(1963) Selective Isolation of Aerobic Actinomycetes. *Appl.Microbiol.* 11: 75-77.

Pikovskaya R.I. (1948) Mobilization of phosphorus in soil connection with the vital activity of some microbial species. *Microbiologiya*. 17: 362-370.

Skidmore A.M. Dikinson,C. H.(1976) *Trans.Br. Mycol.Soc.* 66: 57-64.

Stackebrandt, E., Liesack, W. Witt, D. (1992). Ribosomal RNA and rDNA and sequence analyses. *Gene* 115: 255-260.

Warcup, J.H.,(1955)The origin of colonies of fungi on soil dilution plates. *Trans. Br.Mycol.Soc.*,38: 298.

Williams, S. T., Goodfellow, M., Alderson, G., Wellington, E. M. H., Sneath, P. H. A. & Sackin, M. J. (1983). Numerical classification of *Streptomyces* and related genera. *J Gen Microbiol* 129: 1743-1813.

Yuan ,W.M.Crawford, D.L. (1995) Characterisation of *Streptomyces lydicus* WYEC108 as a potential Biocontrol agent against fungal root rot and seed rot. *Appl. Environ.Microbiol.* 61(8): 3119-3128.

□□□

Chapter 21

DETECTION OF TOMATO LEAF CURL VIRUS IN CULTIVATED VARIETIES OF TOMATO AND OTHER PLANTS OF SUB-HIMALAYAN WEST BENGAL

B. SAHA[1], D.SAHA[2] AND A. SAHA[1]

[1] Molecular Plant Pathology and Fungal Biotechnology Laboratory, Department of Botany, University of North Bengal ; [2] Plant Biotechnology Labotary, Deparment. of Biotechnology, University of North Bengal

Introduction

Geminiviruses constitute the largest, most diverse and economically important family of plant DNA viruses (Rojas *et al.*, 2005). They infect a broad range of plants and cause devastating crop diseases, particularly in tropical and subtropical regions of world (Mansoor *et al.*, 2006). Geminiviruses are characterized by twinned (Geminate) quasi isometric particles measuring approximately 20nm X 30nm, single-stranded DNA genomes that display high levels of genetic variability (Seal *et al.*, 2006). The family Geminiviridae is classified into four genera, *Begomovirus, Curtovirus, Topocuvirus* and *Mastrevirus*, based on their genome organization, host range and insect vectors (Faquet *et al.*, 2003). The largest genus of Geminivirus family is *Begomovirus*. *Begomovirus* have one or two genome components (designated as DNA-A and DNA-B) which infect dicots and are transmitted by *Bemisia tabaci*. A significant increase in the frequency and severity of *Begomovirus* diseases over the past 20 years have been well documented by Rashed (2006). *Tomato leaf curl virus* (ToLCV) is a whitefly (*Bemisia tabaci*) transmitted Geminivirus (family Geminiviridae, genus *Begomovirus*) causing destructive disease of tomato in many parts of India (Sastry and Singh, 1973 ; Vasudeva and Samraj, 1984 ; Saikia and Muniyappa, 1989; Harrison *et al.*, 1991;

Ramappa *et al.*, 1998; Paximadis *et al.*, 2001, Kirthi *et al.*, 2002, Reddy *et al.*, 2005).

The DNA-A component of *Begomovirus* contains four genes (AC1, AC2, AC3 and AC4) on their c-sense strand and two genes (AV1 and AV2) on their v-sense strand. AC1 encodes replication initiator protein (Rep) (Etessami *et al.*, 1991), AC2 gene encodes transcription activator protein (TrAP) (Hartiz *et al.*, 1999), AC3 encodes Replication enhancer protein (REn) (Elmer *et al.*, 1988) and AC4 gene encodes C4 protein. AV1 gene encodes coat protein and AV2 encodes pre-coat protein. Similarly, the DNA-B component contains two ORFs (BC1 and BV1) which encode for nuclear shuttle protein (NSP) and movement protein (MP) that assist in intra-cellular and inter-cellular viral movement respectively (Lazorawitz, 1999). The invariant TAATATT!AC sequence, located in the intergenic regions, contains the initiation site (!) of rolling circle DNA replication (Fig 1).

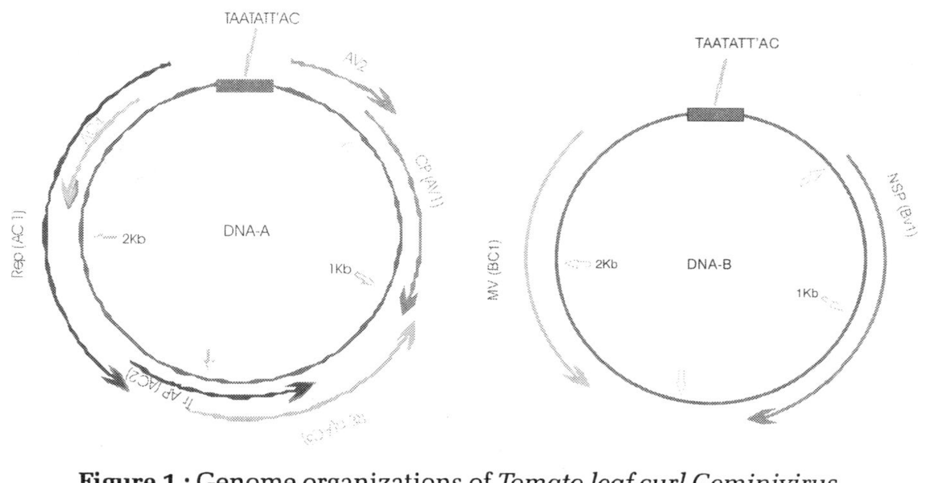

Figure 1 : Genome organizations of *Tomato leaf curl Geminivirus.*

The present study has been undertaken i) to screen the presence of *Begomovirus* in the cultivated tomato plants of different places of sub Himalayan West Bengal by DNA sequence homology, ii) To screen host range of the virus in the associated crops and weeds following PCR amplification and identification of the coat protein gene amplicons and iii) To detect the characteristic geminate virus particles by transmission electron microscopic studies.

Material and Methods

Sample for experiments : Plant materials with viral symptoms (severe leaf curl, shrinking of leaves, yellowing of the veins and stunted growth) of tomato (*Lycopersicon esculentum* Mill.), potato (*Solanum tuberosum* L.) , *Jatropha curcas* L., *Datura stramonium* L., pumpkin (*Cucurbita maxima* Duch), bottle gourd [*Lagenaria siceraria* (Molina) Standl.], *Ageratum conyzoides* L. and *Croton bonplandianum* Baill and *Ricinus communis* L. (Fig. 2) were collected from the crop growing fields of sub-Himalayan west Bengal (which includes Coochbehar, Haldibari and Siliguri). Total 50 samples were collected from the different regions of sub-Himalayan West Bengal.

Leaves of the infected plants were cut with sterile blade to avoid contamination. Cut leaves were poured in zip packs and marked according to the area and date. Leaves were then washed with sterile distilled water to avoid any trace of dust. Similarly, healthy leaves were also collected. Washed leaves were dried with blotting paper and weighed. Dried leaves were packed in a fresh air tight plastic pack and stored at -20°C for further experiments.

Preparation of sample and Electron microscopy. Infected leaves from tomato plants were grind with 0.1M phosphate buffer pH 7.2 in a mortar & pestle. Initially a small amount of buffer was poured in the mortar & pestle and the leaves were crushed to paste. Finally buffer (1ml/g of fresh tissue) was added to the tissue homogenate and mixed thoroughly. One drop of the resultant homogenate was mixed with a drop of 2% PTA (pH 7.0). Twenty microlitre drop of the mixture was placed on a copper coated carbon grid for 30 sec. The drop was socked with a filter paper strip and the grid was allowed to air dry. The prepared grid was examined under JEOL JEM-2010 electron microscope Indian Association for Cultivation of Science, Kolkata and photographs were taken.

Extraction of total DNA of virus infected plants : Total DNA were extracted from the Infected and healthy plants. Cetyl trimethyl ammonium bromide (CTAB) method (Dellaporta *et al.*, 1983) modified by Sharma *et al.*, 2003 was followed for extraction of DNA. One gram of plant tissue was submerged in 5 ml of alcohol for 30 min. and was allowed to evaporate. CTAB-DNA extraction buffer was prepared by

Figure 2 : Naturally infected plants (A. *Datura stramonium*, B. Pumpkin, C. *Jatropha curcas*, D. Tomato, E. Potato and F. *Ricinus communis*) showing typical leaf curl and mosaic symptoms.

mixing Tris-100mM (pH8), 20mM EDTA (pH8), 1.4M NaCl, 2% CTAB and 2μl/ml β-mercaptoethanol. The tissue was then homogenized with pre-warmed (at 60°C) 2% CTAB-DNA extraction buffer. The resultant homogenate was incubated at 60°C for 1h in a water bath, occasionally mixing by gentle swirling. After the incubation the homogenate was removed from the water bath and 0.6 volume of chloroform-isoamylalcohol (24:1) was added. The homogenate was mixed thoroughly by inversion of the up side down and again down side up. After 15 minutes the mixture was centrifuged at 15,000 rpm for 10 min. Aqueous phase was transferred to another tube and DNA was precipitated by using 0.6 volume of isopropanol. DNA was pelleted at 12,000 rpm in 1.5 ml centrifuge tubes. DNA was washed with 70% ethyl alcohol and dried overnight. Dried DNA was dissolved in $T_{10}E_1$ buffer (pH 8). RNAse (2.5 μl) was added to 0.5 μl of crude DNA and mixed gently and incubated at 37°C for 1h. After this 0.3 ml of chloroform-isoamylalcohol (24:1) was added and was mixed thoroughly. Finally the resultant mixture was centrifuged for 15 min at 15,000 rpm. Supernatant was removed carefully so that no white interface layer was mixed with the supernatant. DNA was reprecipitated from the supernatant by mixing absolute alcohol (supernatant: Alcohol::1:2). DNA was pelleted by centrifugation and washed with 70% alcohol. Dried DNA was redissolved in $T_{10}E_1$ buffer (pH 8) and stored at -20°C for further use. All extracted DNAs were diluted 10-fold in sterile distilled deionised water before amplification through PCR as suggested by Reddy *et al.*, 2005.

Primers used :Presence of *Begomovirus* specific DNA sequence was screened in the diseased samples, showing leaf curl by using specific primers through PCR amplification technique. The primers as proposed by Reddy *et al.*, (2005) were used in the present study. The DengA and DengB primer pair targets a ~530bp region that start at the DNA-A origin of replication and continues to the 5' end of the AV1 (coat protein) gene. CRV301 and CRC1152, Cp specific primer amplifies the coat protein gene of all mono- and bipartite Indian TLCB DNA-A sequences available in GenBank.

PCR amplification : Polymerase chain reactions (PCRs) were performed in 25 μl reaction mixtures containing 2 μl of template DNA, 0.5 units of *Taq* polymerase (Genie, Bangalore), 1.5mM $MgCl_2$, 0.2mM dNTPs, 1 pmol of each forward and reverse primers in 1X reaction buffer (*Taq* buffer B). The amplification was carried out using a Gene Amp 2400 thermal cycler PCR system (Perkin Elmer). Amplification programme consisted of one initial cycle of denaturation at 94°C for 1 min. annealing at 52°C for 1min 30sec and extension at 72°C for 2min. After the initial cycle, 40 cycles of denaturation at 94°C for 50sec, annealing at 52°C for 45sec and extension at 72°C for 1min 30sec were allowed to perform. After that amplification programme was continued for 10 minutes at 72°C. Amplification products were electrophoresed at 5V/cm through 1.5% (w/v) agarose gels in 1X Tris Acetic acid EDTA (TAE) electrophoresis buffer, and visualized under UV after staining gels for 30min in 1 μg/ml ethidium bromide.

DNA sequencing and phylogenetic analysis : The amplified products of expected size were directly sent for sequencing. The sequencing was done from DNA sequencing service, Genie, Bangalore. The sequence was compared to the equivalent sequences from a range of other Geminiviruses present in sequence bank. Multiple sequence alignment was carried out using the software MEGA4 (Tamura *et al.*, 2007). The evolutionary history was inferred using the UPGMA method (Sneath and Sokal, 1973). The evolutionary distances were computed using the neighbor-joining method (Saitou and Nei, 1987).

Results and Discussions

In the present study during November to January of 2008-2009 severe leaf curl and mosaic symptoms of tomato, potato, *Jatropha curcas*, *Datura stramonium*, pumpkin, bottle gourd, *Ageratum conyzoides* and *Croton bonplandianum* and *Ricinus communis* were observed in sub-Himalayan West Bengal (Fig. 2). In all the fields a substantial number of whitefly populations have been observed. The infected plants showed curly leaves at early stages and then gradually became distorted and puckered.

Morphology of virus particles : Electron microscopic examination (Fig. 3) of crude preparation of ToLCV revealed the presence of isometric, pentagonal and 'paired Gemini virus' structures. The dimension of the virus particles were 20 X 30 nm when negatively stained with 2% phosphotungstic acid (PTA) pH 7.0. The bipartite structures of our study are very much similar with that of characteristic geminate particles of begomovirus as suggested by Czosnek *et al.*, 1988.

PCR amplification : Following extraction of DNA and amplification of the specific zones of interest, the PCR products were allowed to run on agarose gel. The PCR products of the samples amplified by using specific primers were used to determine molecular weight by comparison with standard molecular weight DNA markers. Two different types of primers were used for the purpose. The two primers were 'DengA / DengB' and 'CRV301 / CRC1152'. The expected sizes of the amplified DNAs were ~530bp and ~870bp when

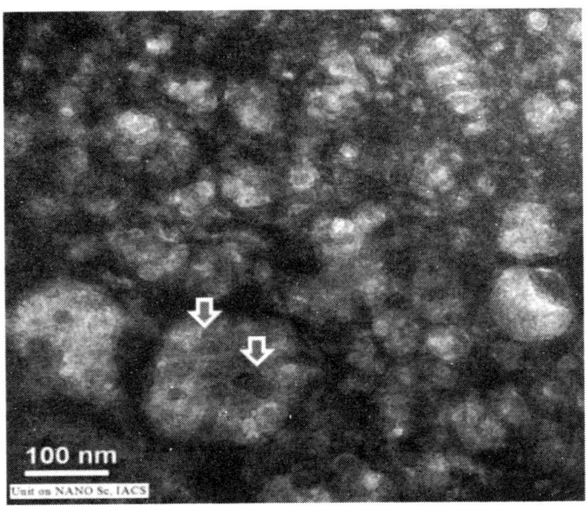

Figure 3 : Electron micrograph of tomato leaf curl virus from infected tomato plant in crude sap preparation. Virus preparation was stained with 2% phosphotungstic acid. Geminate particles are indicated by arrowheads.

'DengA / DengB' and 'CRV301 / CRC1152' primers were used respectively. The results of our experiments when matched with the expected results as described by Reddy *et al.*, 2005, the samples were considered as positive. The PCR products were verified by direct DNA sequencing. The results of some amplified samples along with molecular weight DNA markers have been presented in the Fig. 4. The length of a PCR product of a tomato sample was found to be 480bp after sequencing although Reddy *et al.*, 2005 could sequence up to 530 bp. Eight samples out of fifty samples collected from Coochbehar, Haldibari and Siliguri gave positive results. Samples were assumed positive when they showed bands at 530 bp after amplification in agarose gel. Amplified DNA samples were sequenced and a 480 base pair long partial coat protein gene were observed. The gene starts at the DNA-A origin of replication and continue to the 5' end of the AV1 (coat protein) gene of begomoviruses. The primer pair targets the positive samples were again subjected to another PCR amplification using degenerate "CP primer" to amplify the coat protein gene. The presence of the virus was also confirmed by PCR in two weeds and in six economic plants growing around tomato growing fields of sub Himalayan West Bengal.

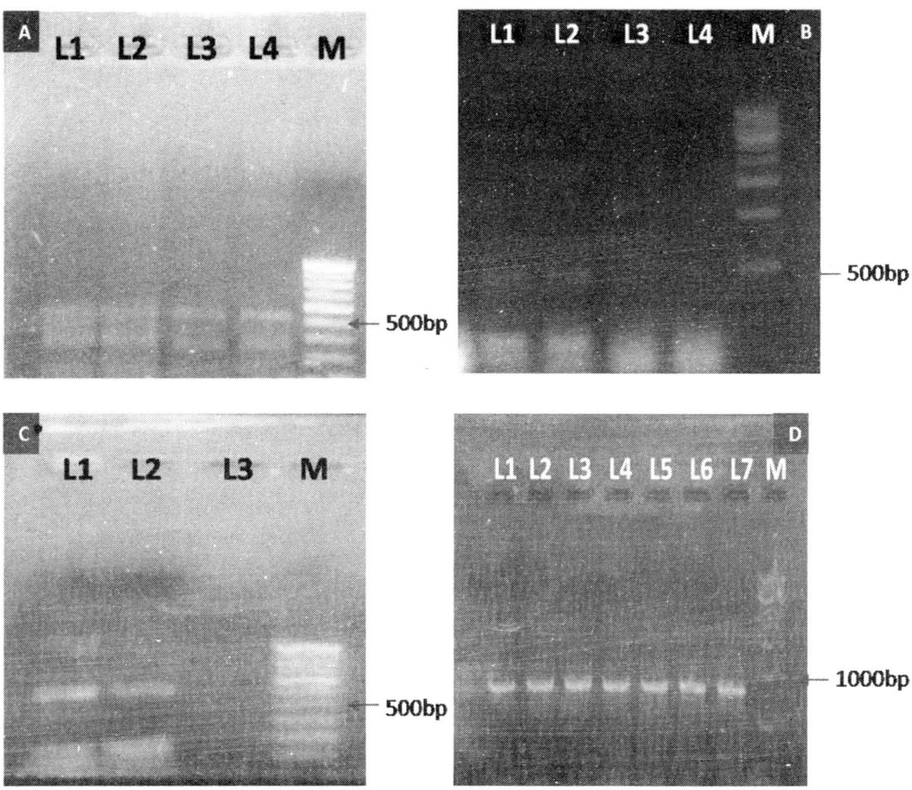

Figure 4: Agarose gel electrophoresis of amplified PCR products using DengA and DengB primer pair, A (L1-Tomato, L2-Potato, L3- *Datura stramonium*, L4-Pumkin, M-Marker), B (L1-*Jatropha curcus*,L2- Bottle gourd, M-Marker) and C (L1-*Ageratum conyzoides*, L2- *Ricinus communis*, M-Marker). Agarose gel electrophoresis of amplified PCR product using "CP specific" primer pair, D (L1-Tomato, L2- Potato, L3- *Datura stramonium*, L4- Pumkin, L5- Bottle gourd, L6- *Jatropha curcus*, L7- *Croton bonplandianum* and M-Marker.

Phylogenetic analysis: BLAST programme has been used to compare the AV2 and AV1 (partial coat protein) sequence from tomato for the homology with sequences available in the GenBank. Amplicon (480 bp) of tomato leaf curl sample from Coochbehar and other sequences available in the Genbank were compared. Highest 95% sequence similarity was observed with tomato geminivirus AV2 gene and partial AV1 gene for coat protein of isolate Y213. In the phylogenetic tree, 480bp sequence from tomato contained segment of AV1 and AV2 open reading frames (which encode for the partial coat protein) formed a cluster with Tomato *Geminivirus* (AM 048873) with bootstrap value of 57% (Fig. 5). Several workers also used the degenerate primers to amplify their geminivirus for phylogenetic analysis (Ding *et al.*, 2007; Rajinimala and Rabindran, 2007; Raj *et al.*, 2001). In the present study out of fifty samples tested, nine samples were amplified by the test primers. The amplified samples were considered as virus affected (Ding *et al.*, 2007; Rajinimala and Rabindran, 2007; Raj *et al.*, 2001). On the basis of electron

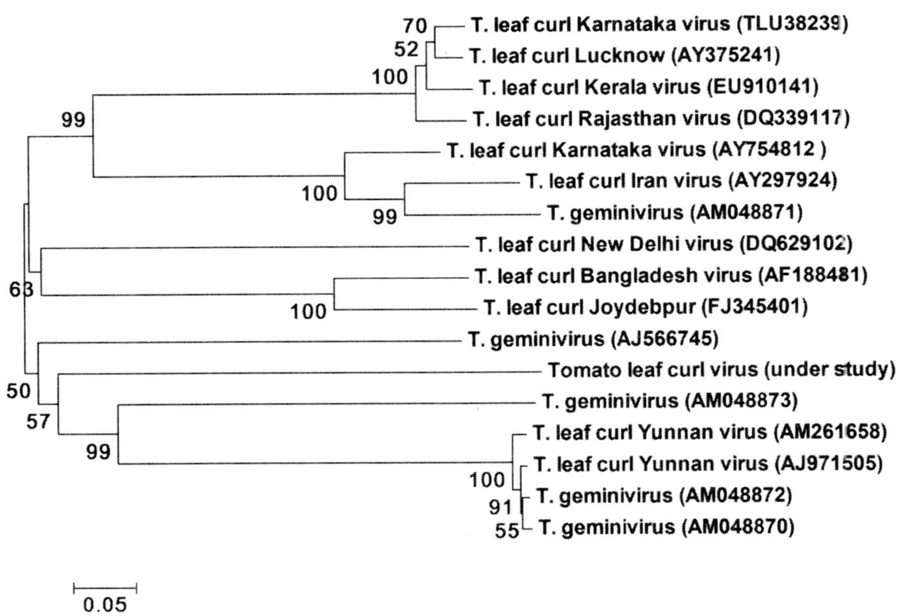

Figure 5 : Most parsimonious tree showing the relationship of 480 bp sequences [Tomato leaf curl virus (under study)] generated from tomato by PCR with primers Deng A and Deng B to the equivalent sequences in GenBank. Numbers at nodes indicates the bootstrap percentage scores out of 1000 replicates.

microphotograph geminate particles were visible. Partial coat protein genes were amplified and sequenced. High sequence identity (86%-95%) and a close phylogenetic relationship with tomato *geminivirus*, the present virus was identified as *Tomato leaf curl virus*. Amplicons of similar size were also observed in potato, *Jatropha curcas*, *Datura stramonium*, pumpkin, bottle gourd, *Ageratum conyzoides*, *Croton bonplandianum* and *Ricinus communis*. Similar host range were also reported by several authors viz. Ding *et al.*, 2007 (in *Datura stramonium*), Wong *et al.*, 1993 (in *Ageratum conyzoides*), Garg *et al.*, 2006 (in potato), Maruthi *et al.*, 2007 (in pumpkin), Reddy *et al.*, 2005 (in *Croton bonplandianum*) Ogbe *et al.*, 2006 (in *Ricinus communis*) and Ranga-Sawamy *et al.*, 2005 (in *Jatropha curcas*).

Several other reports also support the wide host range of the virus. Some of the hosts of the virus are Kenaf (Paul *et al.*, 2006), Bimili jute (Raj *et al.*, 2007), Bun ochre (Chatterjee *et al.*, 2007), *Mimosa invasa* (Koravich *et al.*, 2008), Bitter gourd (Rajinimala and Rabindran, 2007), Papaya (Wu and Zhou, 2006), *Duranta repens* (Tahir *et al.*, 2006), Chilli (Shih *et al.*, 2006), Methi (Raj *et al.*, 2001), common bean (Papayiannis *et al.*, 2007), pepper (Strenger *et al.*, 1990) and bottle gourd (Ito *et al.*, 2008). Electron microscopic study confirmed the geminate particles in twin form and thus also confirmed the presence of the virus in the infected samples (Czosnek *et al.*, 1988).

Acknowledgment

Financial support from UGC, New Delhi is gratefully acknowledged.

References

Chatterjee A, Roy A and Ghosh SK. (2007). First report of begomovirus Associated with yellow vein mosaic disease of *Urena lobata* in India. *Australasian Plant Disease Notes* 2:27-28.

Czosnek H, Ber R, Antignus Y, Cohen S, Navot N and Zamir D. (1988). Isolation of tomato leaf curl virus, a geminivirus. *Phytopathology* 78: 508-512.

Dellaporta SL, Wood J and Hicks JB. (1983). A plant DNA minipreparation: version II. *Plant Mol Biol Rep* 1: 19-21.

Ding M, Luo YQ, Dong JH, Fang Q and Zhang ZK. (2007). First report of *Tomato yellow leaf curl China virus* with DNA β infecting *Datura stramonium* in China. *Australasian Plant Disease Notes*. 2:63.

Elmer JS, Brand L, Sunter G, Gardiner WE, Bisaro DM and Roger SG. (1988). Genetic analysis of tomato golden mosaic virus. II. Requirement for the product of the highly conserved AL1 coding sequence for replication. *Nucleic Acids Res.* 16: 7043-7060.

Ettassami P, Saunders K, Watts J and Stanely J.(1991). Mutation analysis of complementary-sense genes of African cassava mosaic virus DNA. *A J Gen Virol* 72: 1005-1012.

Fauquet EPB, Bisaro DM, Briddon RW, Brown JK, Harrison BD, Rybick EP, Stenger DC and Stanley J. (2003). Revision of taxonomic criteria for species demarcationin the family Geminiviridae and an updated list of begomovirus species. *Arch Virol* 148: 405-421.

Garg ID, Paul-Khurana SM, Kumar S, and Lakra BS. (2001). Association of a geminivirus with potato apical leaf curl in India and its immune-electron microscopic detection. *Journal of Indian Potato Association* 28: 227-232.

Harrison BD, Muniyappa V, Swanson MM and Roberts DJ. (1991). Recognition and differentiation of seven whitefly transmitted Geminiviruses from India and their relationship to *African cassava mosaic* and *Thailand mungbean yellow mosaic* viruses. *Annals of Applied Biology* 118: 299-308.

Hartiz MD, Sunter G and Bisaro DM. (1999). The tomato golden mosaic virus transactivator (TrAP) is a single-stranded DNA and zinc-binding phosphoprotein with an acidic activation domain. *Virology* 263: 1-14.

Ito T, Sharma P, Kittipakorn K and Ikegami M. (2008). Complete nucleotide sequence of a new isolate of tomato leaf curl New Delhi virus infecting cucumber, bottle gourd and muskmelon in Thailand. *Arch Virol* 153: 611-613.

Kirthi N, Maiya SP, Murthy MRN and Savitri H. (2002). Evidence of recombination among the tomato leaf curl virus strains/species from Bangalore, India. *Arc Virol* 147: 225-272.

Koraveih MM, Roff MNM and Othman RY. (2008). First report of whitefly-transmitted geminivirus infecting *Mimosa invasa* in Malaysia. *Australasian Plant Disease Notes* 3:25 -26.

Lazarowitz SG. (1999). Probing plant cell structure with viral movement proteins. *Curr Opin Plant Biol* 2: 332-338.

Mansoor S, Briddon RW, Zafar Y and Stanley J. (2006). Geminiviruses disease complexes: an emerging threat. *Trends Plant Sci* 8:128-134.

Maruthi MN, Rekha AR and Muniyappa V. (2007). Pumpkin yellow vein mosaic disease is caused by two distinct begomoviruses: complete viral sequences and comparative transmission by an indigenous *Bemisia tabaci* and the introduced B-biotype. *EPPO Bulletin* 37: 412–419.

Ogbe FO, Dixon AGO, Hughes JdA, Alabi OJ and Okechukwu R. (2006). Status of Cassava Begomoviruses and Their New Natural Hosts in Nigeria. *Plant Disease* 90: 548-553.

Papayiannis LC, Paraskevopoulos A and Katis NI. (2007). First report of *Tomato leaf curl virus* infecting common bean (*Phaseolus vulgaris*) in Greece. *Plant Disease* 91: 465.

Paul S, Ghosh R, Roy A, Mir JI and Ghosh SK. (2006). Occurrence of a DNA â-containing begomovirus associated with leaf curl disease of Kenaf (*Hibiscus cannabinus* L.) in India. *Australasian Plant Disease Notes* 1: 29-30.

Paximadis M, Muniyappa V and Rey ME C. (2001). A mixture of begomoviruses in leaf curl-affected tobacco in Karnataka, South India. *Ann Appl Biol* 139: 101-109.

Raj SK, Khan MS, Snehi SK and Roy RK. (2007).Yellow vein netting of Bimili jute (*Hibiscus cannabinus* L.) in India caused by a strain of *Tomato yellow leaf curl new Delhi virus* DNA â. *Australasian Plant Disease Notes* 2: 45-47.

Raj SK, Pandey SK, Chandra G, Singh BP and Gupta RK. (2001).First report of geminivirus causing leaf curl disease of *Trigonella corniculata* evidenced by PCR using degenerate primers. *OEPP/EPPO Bulletin* 31: 111-117.

Rajinimala N and Rabindran R. (2007).First report of Indian cassava mosaic virus on bittergourd (Momordica charantia) in Tamilnadu, India. *Australasian Plant Disease Notes* 2:81-82.

Ramappa HK, Muniyappa V and Colvin J. (1998). The contribution of tomato and alternative host plants in leaf curl virus inoculums pressure in different areas of South India. *Ann Appl Biol*133:187-198.

Rangaswamy KT, Raghavendra N, Shankarappa KS, Aswatha Narayana DS, Prameela HA, Govindappa MR, Gururaj Rao MR. (2005). In National Symposium on Crop Disease Management in Dry Land Agriculture and 57th IPS, New Delhi Annual Meeting held at Parbhani, 12–14 January, p. 156 (Abstract)

Rashed MS, Seth LA, Koltunow AM, Randles JW and Rezain MA. (2006). Single-stranded DNA of Tomato leaf curl virus accumulates in the cytoplasm of phloem cells. *Virology* 348: 120-132.

Reddy RVC, Colvin J, Muniappa U and Seal S. (2005). Diversity and distribution of *Begomoviruses* infecting tomato in India. *Arch Virol* 150:845-867.

Rojas MR, Hgen C, Lucas WJ and Gilbertson RL. (2005). Exploiting Chinks in the plant's armor: evolution and emergence of geminiviruses. *Ann Rev Phytopathol* 43:361-394.

Saikia AK and Muniyappa V. (1989). Epediology and control of tomato leaf curl virus in Southern India. *Tropical Agriculture* 66: 350-354.

Saitou N and Nei M. (1987). The neighbor-joining method: A new method for reconstructing phylogenetic trees. *Molecular Biology and Evolution* 4: 406-425.

Sastry KS and Singh SJ. (1973). Assesment of losses in tomato by *Tomato leaf curl virus*. *Ind J Mycol Plant Pathol* 3: 50-54.

Seal SE, Jeger MJ and Van den Bosch F. (2006). Begomovirus evolution and disease management. *Adv Virus Res* 67: 297-316.

Sharma R, Mahila H R, Mohapatra T, Bhargava S C and Sharma M M. (2003). solating plant genomic DNA without liquid nitrogen. *Plant molecular biology Reporter* 21: 43-50.

Shih SL, Tsai WS, Green SK and Sing D. (2006). First report of Tomato leaf curl Joydebpur virus infecting chilli, India. New Disease Reports 13. http://www.bspp.org.uk/ndr/jan2007/2006-65.asp]

Sneath PHA and Sokal RR. (1973). Numerical Taxonomy. Freeman, San Francisco.

Stenger DC, Duffus JE and Villalon B. (1990). Biological and genomic properties of a geminivirus isolated from pepper. *Phytopathology* 80: 704-709.

Tahir M, Haider MS, Shah AH, Rashid N and Saleem F. (2006). First report of bipartite begomovirus associated with leaf curl disease of *Daranta repens* in Pakistan. *Journal of Plant Pathology* 88: 339.

Tamura K, Dudley J, Nei M and Kumar S. (2007). MEGA4: Molecular Evolutionary Genetics Analysis (MEGA) Software Version 4.0. *Mol Biol Evol* 24 : 1596-1599.

Vasudeva RS and Samraj J. (1948). A leaf curl disease of tomato." *Phytopathology* 38: 364-369.

Wong SM, Swanson MM and Harrisn BD. (1993). A geminivirus causing vein yellowing of *Ageratum conizoides* in Singapore. *Plant Pathology* 42 : 137-139.

Wu CZ and Zhou XP. (2006). First report of Malvastrus leaf curl virus infecting papaya. *Journal of Plant Pathology* 88:342.

□□□